Kay Sanson
Liss - Coed
Stratford Rd
Stroud
GL5 4AQ
U.K.

MAJOR MINER

The incredible journey of a
mutinous youth in India
to professor of mining
in Australia

MAJOR MINER

---

*The incredible journey of a
mutinous youth in India
to professor of mining
in Australia*

Gour Sen

PENTAGON PRESS

All names of places & persons are real

Major Miner / **Gour Sen**

ISBN 978-81-8274-618-3

First Published in 2012

Published by

PENTAGON PRESS
206, Peacock Lane, Shahpur Jat,
New Delhi-110049
Phones: 011-64706243, 26491568
Telefax: 011-26490600
email: rajan@pentagonpress.in
website: www.pentagonpress.in

Printed at Aegean Offset Printer Greater Noida (U.P)

*This book is dedicated to
our son Prasanta Kumar Sen whose life was abruptly cut short
at twenty-nine in an accident in Tokyo where he was working
and enjoying a fulfilled life. His position in his Japanese working environment is reflected in this short note presented to us
with his photograph by his employers. The note reads:*

*In memory of Mr P Sen,
You have done an excellent job for us, and have been
loved by all of us, like our brother.
May God bless you forever.
(signed) President, Toshiba Medical Engineering Co. Ltd.*

Preface

It is an ancient Hindu tradition that a personal horoscope scroll (kusthi, in the Bengali language) is written by an astrologer for a child at its birth—at least, for children of families who can afford it—and which predicts the future course of the new-born's life. On it are recorded the time, date and the position of the other planets at the time of birth, and future life events are prophesied to occur at certain time intervals. The owner of the horoscope must not read the contents, and nobody should tell them the prophesies in advance.

I still have my scroll, fragile and half-forgotten among the trappings of a lifetime.

It is fascinating and rather startling for me, a mining engineer living and working over a lifetime in our rational and somewhat sceptical Western culture, to realise that many predictions of the kusthi have been very accurate, especially at critical and even life-threatening moments. The words on the scroll are a thread running through fabric of my life; I will come back to it from time to time in this book.

Acknowledgements

I am indebted to many people for helping with the completion of this book. In particular I would like to thank: my great-niece Adrija Sen for inspiring me to write this book, our daughter Shuki Sen for correcting the syntax errors in the manuscript and my dearest wife, Brigid for suggesting the title of this book and reminding me of various episodes, and for accompanying me on our extensive trips in various parts of the world, and for constantly encouraging me to write this memoir.

I would also like to thank Mike Campbell for his timely assistance and David Rosenberg for his valuable advice.

I wish to acknowledge the professionalism of my editor, Dr Ross Blackwood, and the time and effort he put into smoothing the many rough edges of my original manuscript.

Contents

Preface   7

Acknowledgements   9

PART 1

IN INDIA

Milestones   15

1.   The Early Years   19

PART 2

IN BRITAIN

2.   Cardiff   35

3.   In Newcastle upon Tyne   50

PART 3

BACK IN INDIA

4.   Return to Calcutta   61

5.   Pulling out of India for the Family's Sake   71

PART 4

BACK IN BRITAIN

6.   School Teacher-cum-Consultant   88

7.   Another Turn in the Road   100

8.   A New Opportunity in Australia   106

PART 5

IN AUSTRALIA

9.   Arriving in Sydney   111

10.   Teaching and Research   119

PART 6

TRAVEL TO VARIOUS
PARTS OF THE WORLD

11.   First Sabbatical   125

12.   Facing the Apartheid System in South Africa   128

13.   Rowing in a Pond   135

14.   Sad News from Tokyo   139

15.   First Experiences of China   142

16.   Highlights of our Travels   146

17.   Second Sabbatical   159

18.   Third Sabbatical   174

19.   Promotion and the Co-op Program   182

20.   Fourth Sabbatical   191

21.   Retirement from UNSW   204

22.   New Paths   215

PART 6

RETIREMENT PERIOD

23.   Semi-retired Life   233

24.   Afterwards…   249

PART 1
In India

Milestones

I was born in what is now West Bengal, India—to be precise, at 3 Amherst Way, North Calcutta—into a wealthy Hindu family. My father already had a family by his first wife: seven children, four boys and three girls. After she died, he remarried, this time to the young woman who was to be my mother. My seven half-brothers and -sisters were older than her, and even some of *their* children were older than my brothers and me. I was the youngest of the three boys in my father's second family: Netai, my eldest brother, was five years older than me; sadly, he died of typhoid when I was only three years old, so I have no real memory of him. Dulal, my other brother, was four years older than me.

Photo 1: Gour (left, nine yrs.) and Dulal

I never discovered much about my grandfathers on either my mother's or father's side of the family, but I do recall that a brother of my father's first wife was an alcoholic, and apparently he was what was dramatically called in those days a "shady character" whom my father more or less disowned and kept away from our family.

My father was born in the 1860s into a poor family; he had little formal education but was extremely ambitious. He learnt English, essential if he were to work in a British company and get on in the world. His first such job was at the bottom of the ladder, so to speak, with G. Atherton & Co., a British trading company based in Liverpool, UK. Through his hard work and dedication he rose to become their most trusted employee. I recall that our family was once invited for afternoon tea at his employer's house, and his employer even visited my father at home one time when he was ill in bed—these were extraordinary events in the social structure of the British Raj in 1930s India.

Because of his excellent head for business, my father had become very wealthy by the time I was born: he owned several rental properties in Calcutta, and was part-owner of a factory making hats and caps, and another that produced chutneys and jams; he also owned half a dozen houses in Dehri-on-Sone in the state of Bihar, some five hundred and fifty kilometres north-west of Calcutta. Five of these were tenanted, and one he set aside as a holiday house for the family.

Photo 2: In Atherton's garden. Dulal (2nd left), Gour's father and mother (6th & 7th left), Gour (far right)

Photo 3: The courtyard in our new South Calcutta house

The year after Netai died we moved from the North Calcutta house, leaving our half-brothers and -sisters to stay on there. My father had built a new house in South Calcutta for us, his second family, using exotic and costly imported marble, tiles and mosaics—materials that displayed his wealth. He also bought a second car so that both parts of the family would have transport; but he did retain our Nepalese driver, Bahadur—always regarded as part of the family—to be in charge of our new car.

This was in 1932; I was then four. During the school holidays we used to go by train to the holiday house in Dehri-on-Sone. The house was next to a large strip of farmland and close to the Sone River where we often went fishing. The melons and sugar cane growing in the farm next door were practically inviting us youngsters to help ourselves, but rather than chase us away the farmer would complain to Father. This was an astute move on the farmer's part, since Father then had no alternative but to pay him for the damage. (Of course we were told off sternly—to put it mildly—by Father.) There are other memories of Dehri, too: I saw my first movie there, shown in a field as part of a tea promotion. It seemed so real to me that I thought the man on the screen running towards me with a cup of tea was going to run me over. I recall the hens that gave us eggs, and the baby chickens; and the well in the grounds of the house, our only water supply, and the fun we had hauling buckets of water for the flower gardens and vegetable plots.

Then one day it all changed with the sudden death of my father.

****

From early on, I had ambitions to become famous but now there were no mentors to guide me in the right direction. My young, whimsical mind wandered from one idea to another. My first venture was setting up a children's library; later I aspired to be a writer, then a publisher, and so on. All of this required money, which my mother provided from her meagre funds. This support could be regarded as "spoiling", but in hindsight I acknowledge that it was a magnanimous act by her towards a fatherless boy, barely a teenager. Without her support during that critical period I may well have become a destructive hooligan; for this reason I have a great respect and deep gratitude towards my mother.

During my late teenage years I was fortunate to meet Jayanta Moitra, who started out as my mentor in a rowing club and later became my closest friend. It was terrific to at last have someone I trusted whom I could discuss private matters with. He gave me many valuable suggestions and good advice, and also stood by me in difficult situations. I am deeply grateful to Jayanta for giving me a sense of direction at a difficult period of my life.

***

Later, as an adult I married Brigid, whose background is completely different from mine: English, Christian and from a Western culture. There were (and still occasionally are) clashes between our two strong personalities with such different backgrounds, but I gratefully acknowledge Brigid's enormous capacity for calmly managing our sometimes major day-to-day crises—and for having kept our family life on an even keel for more than five decades. I also thank her for helping me learn to appreciate art and music, which are now a major part of this miner's life.

Photo 4: Jayanta (left) and Gour relaxing after a row in Lake Club garden

1

The Early Years

My father's sudden death caused a traumatic change to our lifestyle. Before his death my father made a Will in which he divided his several properties, dotted around Calcutta, fairly between two sides of the Sen family. After my father's death our half siblings lodged a court case challenging the authenticity of the Will. They tried to prove that my mother was not married to my father and she was only a 'kept' woman, and that therefore all our inheritance was invalid. In those days marriages were not registered. Proof of marriage was dependent on the evidence of those who had witnessed the marriage ceremony. Since it was his second wedding my father had not had an elaborate ceremony and the only guests present were our half siblings and their families. Consequently there were no willing witnesses available to speak for us and our solicitor advised us not to fight the case because it would have been very costly. Since we did not have sufficient funds he advised us to settle out of court. This meant we ultimately had to surrender all the properties except our own dwelling house. Our half-siblings misappropriated most of our rightful wealth in this way and overnight our very comfortable way of life was transformed into a very frugal existence. It seems that my father's brother-in-law—the "shady character"—had reappeared on the scene and incited our half-brothers to do us out of as much of our inheritance as they could. We even lost the car, which was at their North Calcutta house at the time.

Bahadur, our driver, quickly saw what they were up to and refused to be associated with them. He chose to stay with us instead, even though he knew that my mother could not afford to pay him. He did odd jobs for us, and we thought of him as one of the family; this went on for some time until, one day, even he suddenly vanished from our lives.

Our half-siblings' large family was now completely alienated from us. We suddenly found that our side of the family consisted of just four people: our maternal grandmother, Mother, Dulal who was then fifteen, and me, aged eleven. There were no adult men on our side of the family now, and on the other side they were taking advantage of the fact for their own purposes.

In theory, we still owned three rental properties and the half-share in the condiments factory, as well as the house we lived in. Luckily—or cleverly—my grandmother had hidden away my mother's large collection of jewelry, but Mother had to sell valuable pieces from time to time to pay day-to-day expenses.

From a household of about ten servants we could now afford only one, so our mother and grandmother both had to do many of the domestic chores themselves, including the cooking. Understanding all this, I became more self-sufficient and learnt how to darn clothing and to iron and so on, which turned out to be a positive result for me and stood me in good stead later in life.

I was very bitter towards my half-brothers for putting us into this difficult position—so much so that I often invented imaginary plots to kill them. I became and something of a problem child, and quite wild. I not only behaved abominably at home, I was just as bad at school, so objectionable that I was not allowed to continue at any one school for very long. In hindsight I am ashamed for having created so much trouble around me in my early days, and I am very grateful to my mother who tolerated my behaviour and calmly nurtured me through that difficult period.

Dulal was not particularly studious, and he decided to leave school at fifteen to manage whatever properties we had been able to keep. He also tried to get involved with the condiments factory, as the other partner was managing to show a zero profit, but this initiative did not end well.

As for me, I think because of the lack of an adult male authority figure in the family, I became difficult to control. My brother, being older than me, was my de facto guardian but I would not accept this, and often rebelled. I had a furious temper and did all kinds of stupid things when I didn't get my own way: once, to everyone's horror—and mainly to terrify my family—I jumped on the safety net strung between the balconies on the third floor and tight roped all the way across.

At that time, every area of Calcutta had its gang of teenagers. I was the gang leader in our area, having been selected by wrestling matches between would-be leaders. Disputes between gangs were usually settled by a wrestling match, but occasionally by much more vicious fighting. Sometimes I ran away from home. I went to see my very first full-length film at a cinema without asking permission: I even remember that it was called *Tarzan Finds a Son*.

I remember a number of atrocious incidents that I'm ashamed of; one was especially bad—at least, by the standards of the day. One of the teachers at my school was excessively strict. Sometimes, for no reason—to our mind, at least—he would send one of us to the headmaster to be caned; so one day some of the more daring boys (I was one of them) decided to "punish" him. Let me set the scene first: our desks were arranged in two groups separated by a central aisle. The teacher walked up and down the aisle, talking and watching our note-taking skills. In those days most pupils used fountain pens filled with ink. On this particular day, four of us sitting next to the aisle waited until he turned around in his patrolling, then shook ink from our pens all over the back of his white shirt without his noticing it. As soon as the lesson was over, all four of us made up various excuses and left school for the day. I knew that if I ever went back to that school I would be in for truly heavy punishment—so I had to change to another school!

****

During the Second World War there was a large military base about a mile from home where they often practised anti-aircraft gunnery. Naturally this caused quite a bit of excitement among us children. One morning I went with a friend to the camp, complete with my brand new German-made Agfa box camera and a roll of film, and started taking photos of the guns being fired. Before long two native military personnel intercepted us and took us to a room for questioning, convinced that we were taking photos for an enemy spy-ring. They confiscated the film, took our names and addresses and locked us in a room for more than six hours until a British high official arrived to question us. He scrutinised the report on our background check-up, and eventually he was satisfied that we had no connection to any spies, and freed us at midnight, nearly twelve hours later. Although they told me off in no uncertain manner, my family were really glad to have me back!

****

Near our house there was a canal, a tributary of the Ganges, where we would often go to swim. The canal was tidal and usually clogged with mud, so as a rule the water was fairly shallow. One day when I was in the middle of the canal a sudden rush of tidal water filled the canal, completely submerging me. I couldn't swim, so couldn't escape. I swallowed a lot of water and eventually I was exhausted and lost consciousness and sank to the canal floor. The next thing I remember, I was lying on the bank and someone was pumping water from my mouth by pressing on my stomach:

a neighbour—who incidentally was paraplegic—had rescued me. I remember that terrifying day so vividly that even now it sends a shiver up my spine. This life threatening event is recorded in my horoscope.

****

As a teenager I was branded "obstinate" and "stubborn". However, I was an avid reader of children's adventure books and poetry, and this inspired me to be a writer. At first I wrote poems concerning Nature, and sent them to various magazines hoping for publication—but no joy! I then started to write stories and had one of them published in a magazine called *Shishir*. Then, to ensure that my writing would get into print, I started to publish my own magazine in Bengali called *Age Chalo* ('Go Forward' in English). After three or four issues I realised that the venture wasn't viable, especially because the money I had borrowed from my mother had fizzled out. As a consequence, so did my ambitions of becoming a publisher and a writer!

****

After the end of the Second World War there was a huge surge of movement for independence from British rule, and I got involved with a leftist political party called Forward

Photo 5: In a garden, Darjeeling; (l to r) Dulal, Mother, Gour
(around 1946)

Bloc. I took part in a number of political processions, marching to demand various concessions. Sometimes these marches became violent. I was arrested and locked up overnight three times and each time I had to appear in court the following morning, but each time I was released with a warning to henceforth maintain law and order.

Eventually I was elected to be cell leader of a "students' congress". This involved me in arranging strike action in schools and colleges. I held the student meetings in our house, but I discovered that, even then, the meetings were often infiltrated by police informers. Eventually I realised that all our strategies were known to the authorities and therefore all our missions were doomed to failure. This frustrated me, and my interest in politics gradually dwindled. (Incidentally, this was in 1946, around the time my mother and brother took me for a holiday in Darjeeling in the Himalayas, shown in the photograph. As a widow, my mother is dressed in the customary white sari).

****

After I matriculated from school, I began to study law at Vidyasagar College. Near our home was an artificial lake where four rowing clubs were based. I became interested in joining the Calcutta University Rowing Club (CURC), which was a student organisation.

Photo 6: Entrance of the Calcutta University Rowing Club

However, I still couldn't swim, so I couldn't join; but I was so keen to take up rowing that I learnt to swim the required forty metres (about the distance to the shore in the case of a boat capsizing). I became serious about rowing and practised sculling almost every day.

One day I was approached by my college captain, Jayanta Moitra, who asked me if I would be interested in representing Vidyasagar College at the sculling event in the forthcoming Inter-college Regatta. I accepted wholeheartedly. It would be a challenge for which I needed to have coaching every day for the next three weeks, and I practised every morning and afternoon. Of the five contestants in the race, two sculls— one of them mine—were neck and neck in front, well ahead of the others. I could see that my opponent was trying hard to pull away from me. I was equally determined, and performed to the limit of my strength. Eventually I won by half a canvas (about a metre) in record time, and collapsed afterwards for a minute or two from the effort.

Photo 7: Gour after winning the sculling event

This success boosted my self-confidence, and I began to hobnob with the sporting fraternity. In this way I met Pankaj Gupta, who was a very well-known sports administrator in India at that time, and also the chief sports editor of a popular daily newspaper. He invited me to be his newspaper's rowing correspondent, and of course I readily accepted. From then on I was able to make sure that all the rowing event news appeared in the paper and rowing activities gained a higher and more positive profile. This was a paid position, so at the

Photo 8: Group photo of the CURC 's annual prize giving by W Bengal Governor (sitting 8th from left); Gour (sitting far right)

same time I was earning a little money. The payment depended on the size of the news item, but it was usually enough for me to be largely financially independent of my family.

With the regatta win, my star rose somewhat at the club and my friendship with Jayanta became closer; I could see that there was a clash of personalities between him and Shib Ganguly, Captain of the CURC at that time. Under the rules of the club, Shib could not stand for re-election to the following year's captaincy himself so he nominated one of his friends to stand as a candidate. Jayanta himself was not confident of getting sufficient numbers, so he put my name forward. This was supported by a majority of the voting members, and I became the Captain of the CURC. From then on, rowing became my passion and lifelong hobby: since those early days I have taken the opportunity to row competitively in the UK, Brazil, South Africa, Australia and Japan.

****

In 1947, Britain made the decision to end its colonial rule in India. This event sparked bloody Hindu–Muslim riots all over India on the issue of who would control the government of the newly independent Indian nation. Hundreds of thousands of people died—perhaps as many as a million—and mass relocation of populations from one area to another throughout India affected more than twelve million people.

A distant relative, a lawyer whom I knew only as Mr Mallick, was representing us in the court case against our half-brothers. He and his family were affected by the riots: they were Hindus living in a Muslim area. When the riots broke out, one of his Muslim neighbours kindly provided shelter for them in their house. As the ferocity of the rioting subsided a little, the entire Mallick family moved into our house and occupied one floor that was more or less unused; their rent also went towards helping our family's finances.

The end result of the riots became known as Partition—the division of the old India into Pakistan in the Muslim majority area, and India, where the majority of the people were Hindu. The bitterness between the two religions didn't end there, and erupted periodically for months after Partition; indeed, it still affects relationships between the two nations to this day.

****

The preliminary subjects in my law course were history, economics and geography. After two years in the course, I decided that I didn't like the idea of law as a profession but had no idea what I did like. I was in this state of limbo when a sort of miracle happened while I was on a cycle tour with three of my mates to Ranchi, in the state of Bihar, over three hundred and eighty kilometres from Calcutta. (Photo 9)

Photo 9: Cycling from Calcutta to Ranchi with four friends; Gour (top far left)

At one point we noticed a structure that was unfamiliar to us—the head-frame of an underground coal mine, as we discovered. We were curious, and stopped to have a look. We asked at the office if it would be possible to look over the mine, but were told no, all mine visits had to be sanctioned by head office nearly a hundred kilometres away. While we were trying to persuade them to let us have special exemption, the mine manager arrived. Perhaps because he was impressed by the fact that we had cycled such a long way, he gave his permission for a mine visit. We were fitted out with protective gear and lowered two hundred metres down the shaft to the mine workings below. I was fascinated by the fact that a complete working environment could be built so far under the ground, and I was most interested, too, in all sorts of aspects of the mining operation itself (my three mates, however, wanted to get back up to the surface as soon as they could, which is not an unusual reaction, especially for anyone even slightly claustrophobic). This, I realised, was what I had been looking for; it led me to decide that I wanted to be a mining engineer.

However, it proved very difficult to gain admission into a mining engineering school at a university in India at that time.

****

Meanwhile my brother had married Mira, the only daughter of the Mallick family. Some time before that I had met Benu, a college student and dancer, at one of the political party's social events, and we had become friends. In those days, even though I was eighteen years old, such a friendship between teenagers was uncommon in Indian society, and we had to meet in secret. By now, though, with Dulal marrying Mira, Benu was making oblique references to marriage!

One day around that time I was late meeting her at our secret rendezvous, and Benu went to our house to ask where I was. This was a most audacious and daring act on her part; my mother met her and gave her a long talking-to, telling her in effect that her association with me would not go anywhere and that in such circumstances girls bore a black stain on their character forever, whereas nobody ever questioned a boy's past. Benu was very upset by this, naturally enough, and was heading home sobbing when I met her on the road and was able to calm her down.

****

Eventually I was selected for the CURC rowing crew to row at an international regatta in Colombo—but the date clashed with my university finals examination, so I had to choose. I chose rowing and decided not to sit for the exam, much to my family's disgust and strong opposition.

Our team won the coxed fours event, as it turned out— the first time the CURC had done so in that regatta, and a great honour. For my part as a member of the winning team, together with my earlier single sculls triumph at the Inter-college Regatta, the University of Calcutta awarded me a rowing "blue". Photo 10

While in Colombo I met a British participant in the regatta, Lord Craigmyle, who held an executive position in an Indian mining company. Lord Craigmyle provided me with the names and addresses of several universities in the UK that offered mining engineering courses. After returning to Calcutta I applied to all of them. They all refused me, except one: University College, Cardiff, Wales, offered me a place in their course.

Without consulting my family or obtaining their agreement, I decided I would take up the offer and study mining engineering in Cardiff.

Photo 10: Gour awarded Calcutta University "Blue"

This, incidentally, gave me an excuse for a break from Benu, my girlfriend, for a while.

I knew that my family would not support this latest project financially; and, recalling the failure of all my previous whims, I really cannot blame them. My very dear friend, Jayanta, however, knew that my decision was genuine and he agreed to lend me some money. He also had good contacts for arranging my trip to the UK. Das Gupta, another rowing colleague whose family were involved in the airline industry, arranged my flight.

My next problem was obtaining a passport without my family's knowledge. Since there was no system of birth certification back when I was born, I had no document confirming my date of birth, so I invented a birth date: 1st March, 1930—but without a birth certificate, my passport application would have to be countersigned by a senior government official. Luckily, as captain of the rowing club I had become friendly with the local police commissioner and managed to obtain my passport with his help.

The next issue was the need to relinquish all my rights over my assets in India so that my brother could undertake legal transactions without needing my consent. Yet another one of my rowing club friends, Amar Mallick, came to my rescue and arranged the appropriate legal documentation through his uncle.

After making all these covering arrangements, in September 1952 I left home without telling my brother. I simply left a note for him so that he would understand my intentions. I had given my mother some hints, but she didn't take them very seriously and thought that it was just another one of my "running away from home" episodes.

I did tell Benu about my trip to the UK, though, causing a little drama with her on the evening before my departure. We met in 'The Dhakuria Lake' garden area and were sitting in a dark corner having an intimate conversation when a security man approached us. In those days, couples were prohibited from cuddling in public, so we were required to report to his superior in the office. It so happened that I knew the officer in question, also through our rowing club connection. He spoke to me in English so that the security man couldn't follow the conversation, then told us in Bengali that our behaviour was not offensive—instead, he gave the security man a warning, and we were allowed to leave.

PART 2
In Britain

2
Cardiff

The propeller-driven airliner from Calcutta to London did not travel at night, and a journey that today would take perhaps ten hours in those days included two overnight stopovers—first in Karachi, then Athens. At that time relations between India and Pakistan were particularly strained; in Karachi, everyone holding an Indian passport was told to remain at the airport, whereas the other passengers were allowed to spend the night in town. We were herded into rooms in an airport building, with an armed Pakistani military sentry posted at the entrance to make sure we weren't tempted to stray. We were getting pretty hungry by the time we were summoned to the dining room, where I found that the main course on offer was beef. I had never eaten it before, being prohibited for Hindus, but I was so hungry that I ate everything just to survive.

On the next stopover in Athens the following night we were taken to a pleasant hotel in the middle of the city. I was more than impressed with the Western furnishings in my room, and at dinner I was seated at a table for four along with a European couple and their child. After the first course they moved to another table without explanation but obviously feeling unhappy about something. I didn't understand why; it bothered me for months until eventually it dawned on me that, since I wasn't used to the Western style of eating—using cutlery—my table manners must have seemed so obnoxious they couldn't take any more!

As we approached London Heathrow I could see the sloping roofs of the houses, vaguely resembling churches. My immediate thought was that London must have thousands of churches and the people be very religious. This, like so many other first impressions of life in the West, turned out to be rather far from the truth.

In London all the Indian students had to report to the Indian High Commissioner's office to enrol and give details, and some of my loneliness evaporated when I saw so many Indians in that office. My journey overseas is another event of my life that was forecast in my horoscope.

From London I went by train to Cardiff in South Wales and to my digs in Cathedral Road, where the landlady and landlord welcomed me. There were three other students living there. I was to share a large room with Nikko, an Indian student from Uganda. He and the others had settled in a few days earlier.

Soon afterwards I met Mr Morgan, the officer looking after overseas students' welfare. He advised me about the various protocols of university life, and also suggested that I visit the British Council, which was at that time the meeting point for all foreigners living temporarily in South Wales. At times the British Council arranged social functions including, on Saturdays, ballroom dancing; to some extent most of the foreigners treated the British Council as a social club.

****

Soon after I arrived at Cardiff University I went for the usual 'new entrant' interview with Professor John Sinclair, at that time Head of Mining Engineering.

He asked me how much I knew about science subjects. I told him that I had been doing an Arts course in India, so my knowledge of science subjects was nil, except elementary mathematics. Somewhat taken aback, he asked me about Charles' law and Boyle's law.

When he saw that I had never even heard of these things, Sinclair obtained the application form I had sent from India—and saw that I had been admitted into the mining course by mistake!

Since I had no science education background, he explained that obviously I could not contemplate an engineering course, and had only two options: either enroll in Arts, or take a one-year foundation course in science subjects at a technical college and then reapply to the University. I was of course dismayed at this news. I flatly refused to take up another Arts course after all the trouble getting to the UK. Sinclair then rang a Dr Evans, Principal of Treforest Technical College (now the main campus of the University of Glamorgan) and explained my situation.

Dr Evans gave me most valuable advice: university places are very competitive, he said, so don't let it lapse—spend a year studying the lacking science subjects privately. He said that since the university had accepted me, my place was secure; what is more, students were allowed to take two years to complete first-year Engineering.

My strategy would be to attend all the lectures—even though I wouldn't understand them—and sit the examinations at the end of the year, knowing that my marks would be abysmally low. Meanwhile, I would arrange for postgraduate students at the university to give me private tuition in Physics, Chemistry and Mathematics. The only other first-year Engineering subject was Engineering Drawing, which was new to the other first-year students as well. It was the only subject I passed at the end of my first year at Cardiff.

Accident during a Chemistry Experiment

In the first year of my degree course I was in the Chemistry laboratory one day, attending a practical class. My task was to heat a compound of liquids over a bunsen burner. It was on a Wednesday, and our chemistry practical was always followed by sports activities. When I was standing over the bunsen burner, waiting for the experiment to end, the bell went for the end of that session before the experiment had been completed, so, being keen to get onto the sports field, I thought I would speed matters up by giving the test tube a little shake. Immediately, the liquid compound seemed to explode and some of the contents shot up into my eye. I was in real agony and could not see.

I was rushed to hospital where I was very fortunate, as the eye surgeon happened to be on the spot that day so he examined me right away. I was facing the nightmare prospect of losing my eye-sight. While I was 'lying there, I overheard a telephone conversation where someone was enquiring about my condition. The response to a question went something like this:

> "Yes, we have treated his eyes. The right eye looks more hopeful than left one. We shall find out for sure whether his sight is affected tomorrow when the bandages are removed" —a very scary conversation to overhear!

The doctors bandaged my eyes with heavily medicated thick padding. On the following day when the bandages were taken off and I opened my eyes, I couldn't get anything into focus and was quite panic stricken. However, when I told the doctors I couldn't see properly they said it was quite normal. My eyes had to be bathed to remove the residual medication, and then I managed to focus on things. Although the left eye took long time to recover, I was jolly grateful to get my eyesight back, and now I wonder whether this event, which might have led to a major trauma in my life, is recorded in my horoscope, my Kusthi, as it is known in the Bengali language. [see Preface]

In the Engineering course, the first-year was common to all students before they started to specialise in their various branches of engineering. A Civil Engineering student, Alistair Grinstead, one day invited me to have dinner with his parents. This turned out to be a momentous occasion. Alistair's mother, Minnie Grinstead, immediately adopted me as a member of the family. I had tremendous support from the Grinstead family throughout my stay in Cardiff and beyond. I have kept in touch with the family, and am still in touch with Alistair to this day.

****

After that I grew more accustomed to the Western way of life, and realised that I had to adjust myself to my surroundings. For instance, I worked as a sports announcer for the BBC World Service to cover sporting events in Bengali, including the 100th Oxford–Cambridge Boat Race in 1954 (which Oxford won). For this, the BBC paid me approximately one guinea (£1/1/-) per minute of air time. (Document 1)

In mining engineering courses it is mandatory for students to gain practical experience in mines during every university vacation, usually working as a labourer in various parts of the mine to see as many facets of the mining operation as possible. I had never worked as a labourer before, so I was somewhat apprehensive and managed to avoid it in the first vacation period between my two first-year courses. Well before the second vacation period, however, Professor Sinclair personally called me and told me that my training had been organised to take place at an underground coal mine— Tirpentwys Colliery, near Pontypool in South Wales—and that I was to report to the colliery manager, George Griffin, on the Saturday before I started work.

I duly arrived at the colliery on the Saturday morning and went to the manager's office. The door was closed and when I knocked, a booming voice reverberated, 'COME IN'. When I entered the room I saw a large man sitting behind an impressive executive desk. He inspected me from head to toe a couple of times, then told me to be seated. The interview went something like this:

TALKS (Live or Recorded)

THE BRITISH BROADCASTING CORPORATION

Head Office : BROADCASTING HOUSE, LONDON, W.1

Broadcasting House, 38-40, Park Place, Cardiff

TELEPHONE AND TELEGRAMS : CARDIFF ~~?~~26231.

Our Reference :   IE/03/T   15th April, 1954. .... 19......

DEAR   Sir,

We invite you to prepare and deliver a talk(s) as detailed below, for broadcasting or for recording for subsequent broadcast reproductions, upon the conditions printed overleaf. If you accept, kindly sign and return the attached confirmation sheet, or reply otherwise, as soon as possible. (See condition I overleaf.)

Service.. Eastern/Indian/Bengali.

Date of Recording.................................. 7th April, 1954.

Time of Recording................................. 5.30 p.m.

Place.. Bristol.

Date of First Broadcast........................... 10th April, 1954.

Time of First Broadcast........................... About 5 mins. between: 2.15 - 2.45 p.m.

Title.. 100th BOAT RACE: SCRIPT AND READING.

Fee... £5 5 0d. (Five guineas) Plus Rail Fare

Cardiff/Bristol 11/6d. (Eleven shillings and sixpence)

Letters addressed to speakers c/o the BBC will be forwarded, but for statistical purposes the letters may be opened before being forwarded unless we are notified of any objection. Letters marked " Personal " are forwarded unopened.

G. Sen Esq.,
29 Richmond Road,
Cardiff.

Yours faithfully,
THE BRITISH BROADCASTING CORPORATION,

Welsh Programme Executive

KR
P/644/W/P 11-12-52 3,000

Document 1: Pay slip from BBC after my radio talk on Oxford-Cambridge Boat Race

Manager: Have you worked before?

Me: Of course I have.

Manager: What kind of work?

Me: The work involved in studying very hard for examinations.

Manager (showing his biceps): That's not WORK! I mean work with your muscles, man!

Me: Sorry—I haven't as yet.

Manager: Right. First two weeks will be night shift. Report to the overman[1] at 9.30 pm tomorrow, and he'll give you your instructions.

I was staying not far from the colliery in Pontypool with the family of another student (Ken Watkins) as paying guest. His mother prepared a pack of cheese and onion sandwiches for my crib, or meal. With this I took my work clothes, boots and a towel on the colliery bus, arriving shortly before nine-thirty, changed in the communal bathroom, and reported to the overman as I had been told. He then introduced me to my workmate for the next two weeks. Every worker was given a tag with a unique number. This tag was to be put on a hook at the lamp cabin—where the cap lamps were supplied—before going underground, and taken off when the shift was over, to ensure that everyone was accounted for. I was given a comprehensive safety familiarisation exercise before I was allowed to go underground. Since this colliery potentially had methane gas emissions, smoking was strictly prohibited, so everyone was searched for matches or lighter before going underground; habitual smokers had to chew tobacco when underground to ease their nicotine cravings.

****

Steel arches to be used for roof support were loaded into four-wheeled trucks, or tubs, which ran on rail tracks. Several

1. The hierarchy of management in a coal mine in Wales was then, and still is in Australia: manager, under-manager, overman, deputy (and shotfirer, if blasting was used).

tubs were coupled together and pulled by a rope haulage system from the bottom of the vertical shaft, or plat, to a point closer to the working coal face.

My assigned job, with my workmate, was to carry half-sections of these steel arches from the end of the track to the end of the supported section of the haulage road. This work was arduous, and walking back and forth was boring. I was cursing the mine manager for sticking me in this kind of job. At one point during my first night shift I became so despondent that I felt like jacking in the whole thing and taking up another profession—but I had to remind myself of my mission! By the middle of the shift I was staggering with exhaustion. My mate noticed, and told me to take a break in one of the manholes (holes cut in the wall for taking shelter during haulage operations). For safety reasons, sleeping was prohibited anywhere underground and anyone found by a mine official doing this would be blacklisted. Gradually I became used to that rather unusual rhythm of life.

Photo 11: Gour with Mr Griffin (Manager, Tirpentys Colliery

Happily, every few days the manager very wisely transferred me from task to task. I continued at Tirpentwys every vacation after that. Eventually I became so familiar with the mine that, on the occasional Saturday when the colliery operation was fairly straightforward, he left me in charge of his office. Photo 11

****

At the end of the year of my second attempt, I comfortably

completed all my first-year subjects. I also moved to cheaper digs a little further away from the university. At the same time I became involved with the university rowing club, which was still in its infancy in the early 1950s. The club was housed in a small shed on the River Taff and they owned only three old boats. In one corner of the shed was an oil-burning stove to provide some degree of warmth when the outdoor temperatures were close to zero, and this area was allocated for changing into rowing gear. Fortunately—or unfortunately—there were no female rowers at that time.

The rowing section of the river was rather short, as the Taff had a number of weirs, and was fairly fast-flowing. Although we had a stretch of about 1000 m for training, an arched bridge spanned the river at about the middle of the stretch. The arches were not wide enough for any of the boats with oars flanked, so as we approached the bridge the coxswain would shout "oars in", and after passing the bridge we were given the order "oars out"; there were occasional mishaps when the oars were damaged by striking the wall, not surprisingly.

We practised mainly in a Four. When the preparation for the forthcoming Intercollegiate Regatta was looming, our coach (a Cambridge Blue) selected the first three places in the Four. Then he had to choose between me, weighing nine stone (57 kg), and another fellow (about 70 kg), for the bow position in the boat. After trials he chose me—perhaps, due to my previous experience, my technique produced the better results—and the other contender was made cox for all our competition fixtures for one year. This extracurricular activity was a welcome relief from study, and also provided me with social interaction and travel to various parts of Britain for regattas. Photo 12

One of my Indian rowing friends from the Calcutta days, Amar Mallick, was in London for further studies. He visited Cardiff one weekend, and my landlady kindly agreed to provide him accommodation. Later he invited me to stay for

The trial 'A' crew practising at Hereford. Left to right: G. Sen, D.P.Jones, D.J. Bale, R. Calvert, cox, M. Evans.

**Photo 12: Practicing for a Regatta in Hereford;
Note: snow on the ground**

a weekend in London; although Cardiff is less than three hours travel from London, I still had not yet had any chance to see the capital.

So I went to London by train, only to discover the following day that he had severe financial difficulties due to problems obtaining money from India. He gave me a cheque and asked me to draw out some money from his bank account—but, in short, it bounced. Plan B was to borrow some money from a friend of his who lived locally, but wasn't home at the time; by now my small funds had also run out.

By lunchtime we were both very hungry; Amar came up with a plan that I had no option but to follow. We went to a large self-service Lyons Corner House restaurant and sat at a just-vacated table where the previous customers had left quite a lot of uneaten food, already paid for. We gobbled up the leftovers, then sluiced the used cups with tea from the pot before pouring fresh tea in the "clean" cups ... anyway, it allayed our immediate hunger and thirst. Fortunately for me, I had a return ticket, and I left for Cardiff on the train that evening.

My examination results improved in my third year at the university; in fact, I scored joint first in the Mine Surveying paper.

****

In those days the University Students' Union used to organise a 'Saturday Hop' dance, which was a place for the male and female students to socialise.

The girls sat in a row along one side of the dance hall, and the boys sat opposite. When the music started, the boys would get up and walk over to the girls' side to invite his chosen girl to dance. If she wasn't interested in him she could refuse, and the boy would approach another girl—but this situation didn't arise very often, perhaps more from the girl's fear of becoming a "wallflower" than her attraction to the boy.

If they got on well the boy might ask the girl to dance again, perhaps leading to dating, even going steady, as it was called then, if everything went well … and that's how I met Brigid, the girl I was soon to marry; she was also at Cardiff University studying languages.

Brigid and I were married in 1955 in a simple civil ceremony where one of my Indian friends, Robi Mondal, was best man and witness. This marriage to a woman outside my own caste was also forecast in my horoscope (see the Preface). We still treasure the photo of us taken by Robi at Roath Park in Cardiff a few days before the wedding. He also took photos during and after the wedding but unfortunately when he came to develop them, Robi had forgotten what was on the roll of film. He used this roll to demonstrate to a mate how to develop films – and lost the lot. (Photo 13)

Initially I hesitated to tell my family in Calcutta about our wedding, but at the same time I also refrained from asking them for money to support us. My mandatory long vacation work in the collieries gave some financial support in vacations, and Brigid's State Scholarship money was a significant help during term-time.

Photo 13: Shortly before their marriage Brigid and Gour standing on a a bridge, Roath Park, Cardiff

Photo 14: Brigid and Gour holding their recently born first child

After a couple of years, the mine manager George Griffin and I had become good friends. Brigid and I went to dinner one Sunday at his home, which was not far from the mine. After dinner we were so deep in conversation that we lost track of the time and missed the last bus back to Cardiff, so George and his wife Nellie very kindly invited us to stay the night, and even gave us the use of their own bedroom!

****

When our daughter, Anita, was born in August, 1955, I used the opportunity to let my family know that not only was I married, but that we also had a baby. I told them all about Brigid, and sent photos of us with the baby. They replied immediately saying they were happy for us, for which we were much relieved. (Photo 14)

Brigid and I were both busy in our final year looking after our baby as well as cramming for examinations. Little Anita was in day care with some other children on most days for a small fee. Our finances were still rather precarious, and

Photo 15: Brigid and Gour after their graduation (1956)

one of Brigid's lecturers arranged for the university to provide an interest-free loan for baby care. Brigid sometimes had to take the baby with her to lectures, which caused some amusement among her classmates! But we both worked hard for our final exams, and the prize for our efforts was that we graduated on the same day. (Photo 15)

The mid-1960s was the time of the Cold War, and the politicians were gearing up to construct shelters and tunnels in case of a nuclear attack by the Soviet Union. The Auxiliary Ambulance Service was part of this preparation.

There was an appeal for volunteers as Ambulance Service auxiliaries, particularly at weekends and evenings when they were short of professional staff. Since every mining engineer had to have first-aid training and certification, my application was readily accepted.

A volunteer's task was to stand by at the ambulance station and accompany the ambulance driver, who was a certificated paramedic, to help carrying a patient on a stretcher if needed. Sometime these calls came from very seriously ill patients, and we had to deal with them as quickly as possible.

Prior to this I couldn't drive. The head of the ambulance service arranged for me to have driving practice whenever there were sufficient numbers of staff at the station. In this way I was able to get a fair amount of practice by driving an ambulance; in the 1950s these ambulances did not have power steering (despite having been invented some thirty years earlier). Eventually I took the test driving an ambulance, and obtained my driver's licence.

****

After we graduated I took a full-time position in a colliery belonging to the National Coal Board, as it was called at the time. I also enrolled in an external research-based master's degree course. Coincidentally, the research topic was closely related to my work at the mine. After two years' work, I completed writing the thesis and a few of my research articles were published in various technical journals. My first public presentation was at a meeting of the Institute of Mining Engineers (South Wales Division) in Cardiff on 11th December 1958, and it was reported in the local daily newspaper, the *Western Echo*, on 12th December. (Document 2)

The editors of these journals came to know me through these publications, and I became an honorary staff correspondent covering conferences on their behalf, which gave me the opportunity to attend various mining conferences as a non-fee-paying "press reporter", and for which I was paid a small fee.

THE MAN WHO LEARNED ENGLISH

By Our Industrial Correspondent

WELSH engineers praised the feat of an Indian research worker who could not speak a word of English six years ago, but last night talked to them for an hour on a technical mining subject.

When Mr. Gour C. Sen arrived in this country from India in 1952 he could speak only his native language—Bengali.

He came to study mining engineering, but first he had to master a new language.

He did it so well that within four years he graduated from the Mining Department of the University College, Cardiff, with a science degree.

For the next two years the South-western Divisional Coal Board employed him as a research worker.

He made a special study of a highly technical mining subject: strata bolting—a special method of supporting roofs and walls of colliery roadways.

Long journey

And for a detailed thesis he wrote, he was awarded a master of science degree

He left South Wales to become a research associate of King's College, Newcastle. Now he has switched his studies to the salt mines of Cheshire.

But he made the long journey to Cardiff last night to talk to the South Wales Institute of Engineers on strata bolting in the coalfield.

Was it much of an ordeal?

Mr. Sen smiled. "It was all right. I know the subject well, so I just spoke what I knew to be right," he said in broken English.

Document 2: Media report after my technical presentation at a professional meeting

3

In Newcastle upon Tyne

I was offered a Research Associate position at King's College, University of Durham in Newcastle-on-Tyne; in 1963 King's College became The University of Newcastle upon Tyne. This externally funded position allowed me to enrol for a PhD degree. My research program was to re-design the workings of an underground salt mine in Winsford, Cheshire; the mine was owned by Imperial Chemical Industries (ICI)—now part of the Dutch conglomerate AkzoNobel.

I often had to travel to the mine from Newcastle to set up instrumentation to measure the deformation occurring over time in different sections of the mine. Eventually I bought an old van—a little old Ford which we christened Aunty-car because of her stately rate of progress, the first vehicle I had ever owned—for which I could obtain travelling expenses from the university, and which was handy for my family as well.

****

The head of the Mining Engineering Department at King's College at the time, and later at Newcastle University, was Professor Edward L. J. (Ted) Potts, a highly respected figure in the mining industry, well known as an academic and consultant to mining companies in the UK and in countries all over the world—France, South Africa, India and Australia, to name just a few. Potts was also adept at obtaining research funding for projects employing postgraduate research students whose degree programs were then largely related to resolving a problem being experienced by the sponsoring company.

The funding for my own ICI-sponsored project was reviewed each year by a panel of experts to whom the progress of my research was presented. In the two-hour drive from Newcastle to Winsford it was my job to brief the Prof (as he liked to be called) on the progress of the work. He would question me until he was completely in the picture; on some occasions, if the two hours wasn't long enough for me to tell him the full story, we would stay at a hotel and finish it there. The next morning I would sit beside the Prof at the meeting and listen while he recounted the entire story to the ICI representatives as though he himself had done the work, or at least supervised it. Occasionally he might call on my services during the question/answer time to supply any very detailed information.

****

In Paris

When I had nearly completed my research program, a mining-related international conference was about to be held in Paris, and all postgraduate students from our department were encouraged to attend. Six of us took part; some were asked to contribute a paper at the conference in return for having our out-of-pocket expenses heavily subsidised, and I was one of those asked.

After arriving in Paris, we registered at the conference office, then attended a lavish reception, and the champagne flowed freely! Much later at our hotel, I noticed that the bed of my room-mate, Graham Johnson, was empty. The following morning I found him deeply asleep in his bed, his face plastered with mud—he had enjoyed the free champagne (a very rare experience for a student at that time) somewhat beyond his capacity and, after the official reception, had gone to a bar near Montmartre to continue drinking. The next thing he remembered was lying in a gutter in the wee small hours, being prodded awake by a gendarme. Then he realised that he had lost his watch, wallet and passport. Very kindly guided to the hotel by the gendarme, Graham then managed to grab just two hours' sleep in the hotel. Some night!

****

I presented my research findings at the discussion period of one of the conference sessions. I had never been to a nightclub, and mentioned it to a young Frenchman at the conference. I said, "Paris is famous for nightclubs, but can you recommend one?"

"Well, there's the Moulin Rouge, Folies Bergère and so on, but when our French friends come to Paris from the country we tell them to go to Club Zodiac in Rue Saint-Denis. There's a lot of temptation, so don't take too much money. Just take 20 francs, which will cover the entrance fee and one glass of beer. If you order champagne, which will cost you 50 francs, it means you want a girl," he said. This place was, of course, in the middle of the Paris red light district, but I didn't know that.

I followed his advice and took 20 francs plus an extra 5 to cover the metro fare, signed a form to become a temporary member and paid up. There was a big room upstairs with a central stage and a gangway for the performers. Loud pop music was blaring away. Seats were arranged all around the stage and there were curtained cubicles around the walls. I was early, so I sat in the front row close to the stage. Another young delegate from the conference fellow sat next to me; having a companion gave me some comfort. It so happened that nightclubs were new to him as well. The room was soon almost full, mainly men but also a sprinkling of women.

The show started with relatively mild striptease acts, but progressively became very daring, although the lighting tricks made them all seem fairly modest. I noticed that most of the cubicles around the stage were occupied by middle-aged men with girls on their laps, sipping champagne. The show became more explicit, with live sex acts on stage loudly cheered on by the audience. Then I saw a crowd of new customers coming in the door: lo and behold, there was our Professor Potts with some high officials from the conference. Prof and I saw each other and laughed. I said to him, "I don't want to know whether you're going to order beer or champagne, so I'd better leave," and I went back to the hotel. The "Ted Potts and the Paris nightclub" incident became a standing joke at the university afterwards.

****

In 1961 I completed my research program and my thesis was passed by both external and internal examiners—my PhD work was complete. King's College had become The University of Newcastle upon Tyne during this time, so we had the option of attending the degree-awarding ceremony (held in 1962) either in Newcastle or in Durham. The mediaeval-style graduation ceremony for Durham University took place at Durham Cathedral, and together with two other doctoral candidates from King's College, Frank Roxborough and Peter Huchinson, I chose to attend the ceremony in Durham with Brigid and our older children there to witness this momentous occasion. (Photo 16; Photo 17)

Photo 16: Our three children with their mother and granny to attend Gour's PhD degree ceremony; Durham Cathedral (1961)

Photo 17: Gour after receiving PhD degree certificate

After receiving my doctorate certificate I felt elated and
... I think *virtuous* is the word. (Document 3)

****

In Netherlands

I was commissioned by a mining magazine to attend and
write a report of a mining conference to take place in
Scheveningen, a district of The Hague in the Netherlands.
When our friends in Newcastle heard about this they insisted
that Brigid go with me, and they would look after our three
children. The conference was held at the luxurious seafront
Kurhaus Hotel. Most of the delegates were accommodated
there, but we stayed at a private guest house nearby. At
lunchtime, while the other delegates ate at the hotel, Brigid
and I bought snacks from roadside stalls—after first making

UNIVERSITY OF DURHAM

Gour Chand Sen of King's College

HAVING COMPLIED WITH ALL THE CONDITIONS
REQUIRED BY THE UNIVERSITY, HAS BEEN
ADMITTED TO THE DEGREE OF *Doctor of Philosophy*
in Applied Science.

CHANCELLOR

REGISTRAR

DATE 7th July 1961

Document 3: Gour's certificate after completion of PhD degree

sure we took off our conference name tags. This is how we first discovered satay chicken, the dish from the old Dutch East Indies colony that was to become Indonesia. We liked it so much that we often looked for it in restaurants after we returned home to England.

The conference dinner was held in an exclusive restaurant set in a forest, originally the palace of a Malayan sultan that had been dismantled piece by piece and shipped to Holland. We were treated to a lavish Javanese *rijsttafel* (rice table, another Dutch ex-colonial dish), a marathon banquet of more than fifty spicy dishes—another event that will always be fresh in our memory.

****

I continued to work at Newcastle University as a Senior Research Associate in order to tie up the loose ends in the salt mine project, as well as undertaking some other advisory functions and preparing for my First Class Mine Manager's Certificate of Competency examination. I already had the statutory underground mining experience, and it remained only to sit a written examination and oral test. After I had satisfied the examiners I was awarded the certificate. (Document 4)

MINES AND QUARRIES ACT, 1954
(2 & 3 Eliz. 2 c. 70)

FIRST-CLASS CERTIFICATE OF COMPETENCY
(MINES OF COAL, STRATIFIED IRONSTONE, SHALE OR FIRECLAY)

THE MINISTER OF POWER, on the recommendation of the Mining Qualifications

Board and in pursuance of the powers conferred upon him by subsection (1) of section

one hundred and forty-seven of the Mines and Quarries Act, 1954, hereby grants to

... GOUR CHAND SEN ...

of   35, SANDERSON ROAD, NEWCASTLE-UPON-TYNE, 2.

this First-Class Certificate of Competency valid with respect to mines of coal, stratified

ironstone, shale or fireclay.

Dated this   **first**   day of   **February**   , nineteen hundred

and sixty-two.

MINISTRY OF POWER,
LONDON, S.W.1.

An Under Secretary to the
Ministry of Power.

Department of Mineral Resources

Prof. F. F. Roxborough
Head Of School
Mining Engineering
The University of New South Wales
P.O. Box 1
KENSINGTON NSW 2033

C.A.G.A. Centre,
8-18 Bent Street,
Sydney.
Postal address:
GPO Box 5288
Sydney, NSW 2001
Telex AA21708
Our reference: M79/525
Your reference:
For further
information ring: Mr Talbot
Telephone 231 0922
Extension .4317...

Dear Sir   4th March, 1981.

I acknowledge receipt of your letter dated 24th February 1981

and enclose Mr G. C. Sen's United Kingdom Mine Managers

Certificate which has been registered for use in New South Wales.

Other documents submitted are also returned.

Yours faithfully

**Document 4: Mine Manager's certificate from UK which was
endorsed in NSW, Australia**

PART 3

Back in India

4

Return to Calcutta

At the same time I was looking for an opening within the ICI group, it transpired that a Technical Service Engineer was required for ICI's Explosives Division in India. As I hadn't seen my Calcutta family since leaving India nearly ten years before, the position was naturally of interest to me. I would be based at ICI's Indian head office in Calcutta, and another important attraction was that my family and I could therefore live in my late father's spacious house. In due course I was interviewed in London by the manager, Geoff Parkes, and was offered the position.

Shortly afterwards I received confirmation of my job offer, along with descriptions of the responsibilities and privileges attached to the post and the stipulation that I first undergo a three-month training period with the UK's counterpart of Explosives Technical Service. I was to accompany a Technical Service Engineer on various projects as an observer. The experience of becoming familiar with the properties of various commercial explosive products and their safe and efficient use, proved very profitable.

Before we knew it, Head Office was arranging the whole family's trip to India by air, which we didn't feel was appropriate. Firstly it would be a big shock to the children transported so suddenly from a cold country to a very hot climate. Further, we were planning to take with us all our virtually new white goods—washing machine, fridge, cooker, etc.—with us, impossible if we were to travel by air; we were quite adamant about all this. Then we received another proposal suggesting that I travel by air, and the family would follow me by sea. This was also unsatisfactory; we had by that time four children under the age of eight, and a two-week journey with the children in a confined space without any assistance would have affected Brigid's health and mental state. The travel issue almost made me change my mind about accepting the position, but in the end the company reluctantly agreed to provide a sea voyage for the whole family.

In October 1962, the Sen family set sail on a P&O liner MV *Chusan* from Tilbury for Bombay, where we were to disembark; the vessel's ultimate destination was Australia. The highlight of the journey was the ship's passage through the Suez Canal, which took a whole day. The passengers had the option of either staying on board or getting off at the northern end of the canal and rejoining the ship at the southern end, which is what we did. We managed to get a glimpse into the Egyptian way of life, and did all the tourist things—we saw Tutankhamun's mummy in Cairo Museum, entered a the main pyramid, rode on camels. (Photo 18) Another highlight of the voyage was our middle daughter Runa's birthday, which we celebrated with great gusto on the ship.

Photo 18: Our children having camel ride near pyramids

After two weeks of ocean travel, our ship docked in Bombay early one morning, and there the company's representative met us to assist us with the disembarkation formalities. We expected that it would take considerable time to unload our several crates of luggage and pass through Customs. We were taking the train to Calcutta that evening, so we decided that I should stay at the wharf until all our belongings were accounted for and had cleared all Customs' formalities while the rest of the family relaxed at a hotel.

I soon located all our crates, and Customs officers asked me hundreds of questions regarding their origin, use and so on, and insisted on opening each one to inspect the contents. All the while the company rep kept asking me, "Why don't you settle with the Customs officers?" I kept saying, "I'm trying to!" Eventually I twigged—"settle" of course meant "bribe". After more than ten years in England I had forgotten how widespread corruption was in India, including among these Customs staff; anyway, I followed the local custom of doling out bribe money and eventually rejoined my family at the hotel.

Due to some administrative breakdown at ICI, we had been booked in a First-class compartment on the train to Calcutta, not Air-conditioned Class, which would have been more comfortable; what's more, this train was going to follow a much longer route than the usual one, and would take more than two days. Luckily someone also reminded us that bedding was not supplied, and that we had to hire it. When we entered our compartment, we found that several of the facilities weren't functioning—the bathroom was filthy; the sink didn't have a plug, so there was no way we could wash our baby or clothes. To have meals, we had to get off at convenient stations along the way and, after locking up our compartment, go to the dining section of the train. As well as the considerable discomfort all this created for us on such a long journey, most of our children had an upset stomach from the unhygienic food.

Finally our train arrived in Howrah station near Calcutta, where my brother Dulal and other members of my family were waiting to greet us. A staff member from ICI, Pulak Bhose, was also there to meet us. We were then driven to my late father's large house at 3/3A Shahnagar Road where one floor had been designated as our living quarters.

It was quite an emotional experience to meet my mother and other members of my Indian family after more ten years. Brigid and the children felt much relieved that the long journey was over and quickly settled down in the family house. We would be living in the front portion of the four-storey mansion. The two other portions had been subdivided into a number of apartments that were occupied by tenants; the rents from these were the sole income for my brother's family. The lower two floors of the front section were made up mainly of marble, mosaics and Italian ornamental tiles.

****

A couple of days after we arrived I received a phone call from my ex-girlfriend, Benu. How did she know that I was back in Calcutta? God knows! I was somewhat shocked to hear her voice. She wanted to meet me secretly somewhere, a plan which I managed to avoid. We were speaking Bengali, and the conversation lasted less than five minutes, but when Brigid asked me who that was on the phone I lied and said, "My aunty". Although at the time she knew very little Bengali, she was smart enough to figure out that my response was untrue. After a short but rigorous interrogation I spilled the beans—fortunately the girl in question never surfaced again!

****

Living in Calcutta

Although our apartment had its own kitchen, for the first few days our food was supplied from my mother's kitchen upstairs. In the meantime we were busy trying to install all the electrical equipment we had brought from Britain; there was a big problem, however—our items ran on alternating current (AC) whereas the wiring system in the house was direct current (DC). Consequently we had to use an electrical converter so that its outlets provided AC current. However, the converter could not be operated continuously without overheating. Running the fridge in particular was a rather painful operation: the converter had to be switched on and off intermittently to keep the fridge cool. We soon became accustomed to the system, however, and learnt to live with it.

It was not easy for my family to settle down in the Calcutta environment after our fairly smooth life in Britain. There were problems with education for the older children who had to cope with learning Bengali. There a lot of airborne dust which settled everywhere, and the children's health and hygiene needed constant vigilance.

It was not safe or comfortable for Brigid to spend time on her own outdoors in Calcutta, which restricted her lifestyle somewhat. On the lighter side, she learnt from my family of my misdeeds as a teenager and said that, had she known about my murky past, she would never have married me. The family very slowly adjusted to the conditions, however, and began making friends. Our intention was to give it a fair go.

****

ICI had two levels of staff: General Staff and Management Officers. The management staff were subdivided into Groups 1 to 5, depending on their positions. The directors were Group 1, senior managers were Group 2, managers were Group 3, senior staff Group 4 and junior officers Group 5. The higher the group, the higher the salary and the more extra perks. I was recruited as a Group 4 staff in the Commercial Explosives Division. My responsibilities were to sort out clients' complaints and problems, introduce new products into the market, and train industry personnel to use explosives safely and efficiently. Although my office was in the head office in Calcutta I had to spend considerable time on field work or in the factory in Gomia, situated about 300 miles from Calcutta in the state of Bihar.

Shortly after our arrival in Calcutta I was also invited to be a visiting lecturer at the Mining Engineering Department of Bengal Engineering College to teach commercial blasting once a week for a set period. My superiors at the office encouraged this involvement with a local university, which counted as a significant public relations exercise for the company.

Meeting Sir Edmund Hilary

My work took me to various parts of India for discussions with very senior members of industry. One day Mr Keown, one of the directors, called me into his office and introduced me to an extremely tall man, none other than Sir Edmund Hillary—the man who, with the Sherpa Tenzing Norgay, had been the first European to climb Mount Everest some ten years earlier.

His plan was to build a hospital in the Himalayan region for the native Sherpas. His problem was to create sufficient flat ground in the middle of the mountain range to build the hospital and construct a short airstrip for supply planes. He showed us a large number of aerial photos from which he had identified a possible location for the airstrip, but some large boulders could be seen which would have to be removed before any work could be done. There was no vehicle access to that part of the mountain range. All provisions would have to be carried in by porters, which would require huge manpower and would be costly. My job was to estimate the amount of explosives required to shatter those boulders and to brief the team on the blasting procedure. It was not possible to provide an answer on the spot, so I borrowed the photos and took them home to study in detail.

I could see a cow in one of the photographs, providing a useful scale from which I could estimate the volume of the boulders and, from this, the amount of explosives that would be needed. It was imperative that I work out the amount of explosives quite accurately because of the porter access. Unfortunately the company would not allow me to join Hillary's project team, claiming that I could not be spared for such a long time. I trained a member of the team to use explosives safely, but it was a big disappointment to me not to be involved on site in such a project.

****

Short term Military Assignments

In 1962–63, conflict between India and China erupted over border issues; both sides began mustering their military hardware. The Indian Government ordered large companies to nominate staff members who would join the army for a short period of time to assist with this national issue. Working in explosives, I was an obvious choice; my duty was to instruct army personnel how to quickly demolish a bridge to delay the enemy's advance in the event of a retreat by the Indian army. I was seconded to the army on two occasions, once to Sikkim and once to Bhutan, both states on the Himalayan border with China.

Brigid wanted to accompany me but her request was refused, as she was a foreigner and therefore was seen to pose a security risk to the Indian army. In hindsight, this turned out to be the best for her. Our flights from Calcutta to these places were in Defence Department cargo aircraft in which we had to squat on the hard floor during the flights. The sleeping arrangements, also, were very rough and frugal: a hard wooden bench for a bed and a rough woollen blanket into which I had to roll myself like a sausage to get any warmth when sleeping.

During one expedition, I arrived at the airport of the Sikkim capital Gangtok and was met by an army driver with a jeep. He driver told me that our destination was close to the Chinese border, and we would be travelling through heavily forested terrain. During the journey I noticed some decorative objects hanging from some of the trees. I was so curious that I asked the driver to stop the vehicle; he thought I might be obeying a call of nature, so he stopped. I got out and went up close to one of those decorated trees and looked up. I was fascinated by the beauty of those wildflowers hanging from the tree and was admiring them. The driver was puzzled by my attitude. As far as he was concerned, this was a military operation; how could I be so soft as to waste time admiring flowers? Eventually I found out that the flowers I had been admiring were native orchids. This event influenced me to keep and nurture orchid plants later in life.

On another occasion I was engaged in a project training army personnel in the very remote area of Ladakh near Kashmir, in the eastern Himalayas. An army platoon was constructing a road though the rough terrain. There had been some explosives-related incidents, which was the reason for my visit. Explosive cartridges were nicknamed 'sausages'. Normally this outpost obtained their food supplies once a week from a town some two hundred kilometres away. On one occasion the supply convoy was delayed due to a landslide, so the camp commander had to ration food temporarily. One of his men had the bright idea of frying 'sausages' in a frying pan ... that was one of the incidents. I found that army personnel were familiar with munitions but their knowledge of commercial explosives was virtually nil.

I finished my work there in two days, and the convoy from the base camp was not due for another five days to take me back to Calcutta. I persuaded the camp commander to have me driven to Chandigarh, from where I would be able to arrange transport to Delhi and from there to Calcutta. He allocated me a jeep and driver for the overnight journey to Chandigarh. After mess at the camp that night we started out, following a very narrow track, with a high cliff on the left and a drop of about 30 metres into a ravine on the right; almost no straight sections of the track went for more than about 15 metres; and the track itself was only a few centimetres wider than the jeep—so the driver had to proceed slowly and carefully.

I was sitting next to him and kept talking to him so he wouldn't doze off. We started out well, then every half an hour or so he stopped and went to the back of the jeep. I asked him if there was any problem. No, he just wanted to check the tyres. After a while I could smell alcohol, and his driving was becoming erratic. I tactfully—I thought—suggested to him that he might be feeling sleepy, and perhaps I should take over the wheel. He objected strongly, and went on stopping and presumably charging himself up with more booze. At one corner he nearly lost control of the jeep.

I was rigid with the fear of falling over the edge. I realised that if I upset him it would be counter-productive, so I kept on trying to amuse him with stories. After about six hours or so we were on the plain, and the road was wider, with fewer bends. Eventually dawn broke and I was extremely relieved just to know that I was still alive. However, with the tension and without any sleep I was a wreck. Of course, later back in Calcutta I did drop a note to the commander describing the traumatic journey.

****

Obviously most of my professional activities were in the industry sector, such as mining, tunnelling, demolition of structures, road construction and so on. These activities required a vast amount of travelling, and I had to spend a lot of time away from home. As the months went by, although my employer appreciated my dedication, I sensed mounting discontent about it in my family, and my company management team must have noticed a change in my attitude.

One day I was called in to see the managing director who told me that I was being promoted to Group 3. When I read the terms of the new position I realised that I wouldn't be receiving all the privileges enjoyed by other Group 3 staff; however, there were substantial extra benefits involved, including the use of the company guest houses dotted round India. These were like hotels, with meals included, and the staff and families of Group 3 and above could stay there free of charge in their holidays. One of these was in Puri in the state of Orissa, a well-known beach resort on the Indian Ocean.

Taking advantage of my new status, my family and I spent a week's holiday at Puri. What a joyful time for us— the guest house was like a boutique hotel; we were the only people there and they thoroughly pampered us. The children loved the beautiful sandy beach and huge surf as much as Brigid and I did. (Photo 19)

Photo 19: Our children with their mother, Puri Beach, Orissa

5

Pulling out of India for the Family's Sake

The relaxed holiday environment also gave us the chance to discuss our position in some depth. We started to question our mental and physical health. Life was almost dream-like in Puri, but ... in Calcutta?

In Calcutta, Brigid had been housebound most of the time. The constant heat and dust, and the general environment, made her feel that she could not walk about freely in the streets. She often reminisced about her free life in the UK when she used go shopping with the baby in the pram. This lack of freedom was quite a new and difficult experience for any woman used to living in a Western country. The children had had their fair share of illness; although my promotion meant that the company covered the medical expenses of whole family, the children were unhappy when they were unwell. Finally, we questioned the children's education, which was regimented and parochial. All these factors led us to believe that our future was not in India, and that we should pull out sooner rather than later.

After returning from Puri we sorely missed our short holiday. The harsh reality of life in Calcutta seemed more pronounced, and ultimately we decided we must leave

Calcutta. We thought about living in another part of India, and I made some enquiries. Another company, Indian Detonators Ltd based in central India, offered me a higher position as Technical Service Manager; however, this would have meant more upheaval for the children, with a new language and education system. We gave it some thought but decided not to accept, and to leave India altogether. It was not an easy decision. How would our Indian family take it? Where should we go? Since I had only an Indian passport it would be difficult to organise a job overseas.

In the meantime Brigid and I took a couple of weeks' holiday in the north-western part of India: Delhi, Shimla, Kashmir, Jaipur and Agra. In the Clarkes Hotel in Shimla we met a South African-born couple from the UK, Joy and Eric Nell, who we remained friends with for some time after we left Shimla. When we were staying at the hotel the English actress Felicity Kendal was a guest during the shooting of the film *Shakespeare Wallah* (1965), so naturally there was a lot of activity in the foyer most of the time; and in the native Himalayan trees surrounding the hotel lived a colony of monkeys, which were a real pest—your afternoon tea and cakes were not safe from them.

In Kashmir we spent a few days in Srinagar in a houseboat on Dal Lake, and also took a three-day trek across uninhabited territory on horseback from Pahalgam—twenty-one hundred metres above sea level—to visit Kolahoi Glacier, with a guide and a servant/cook and enough food for the three days. The plan was to ride to a rest-house on the first day, sleep overnight there, then, on the following day, visit the glacier and return to the rest-house. After paying the usual fee at the local government office in Pahalgam we borrowed the rest-house key (which was supposed to be available to only one party at a time).

As we approached, we could see from a distance that there was some movement at the rest-house. Sure enough, we found an English couple, Mabel and George Garside, and their party were already occupying part of the rest-house. (After a somewhat frosty start to our relationship, eventually it thawed and we remained good friends until George's death many years later.)

The following morning both parties set off to visit Kolahoi glacier. After a couple of hours trekking uphill we reached a point where the terrain was much too steep for the ponies. This was when we also noticed that Mabel was feeling unwell due to the high altitude (about 2700 metres, nearly 9000 feet) and decided to stay there with the pony keepers while we pushed on to the glacier. The distant view of the glacier was absolutely stunning and awe-inspiring; finally we stood at the foot of the glacier at 3100 metres altitude, where the melting ice is the source of the Lidder River which flows through Pahalgam.

After Kashmir, Brigid and I returned to Calcutta and the Garsides went home to Blackpool. Eventually they emigrated to New Zealand and we kept in contact with them until George's death, after which we lost touch with Mabel, sad to say.

Our flight from Srinagar to Agra via Delhi was delayed due to some technical fault. We were very concerned that if we missed the connecting flight from Delhi our itinerary would be spoilt. At my request, an official at Srinagar Airport phoned Delhi and requested them to hold the Agra flight for us—and when our plane touched down in Delhi, a vehicle was waiting to ferry us from one plane to the next. I have never seen such an efficient transfer in my life, before or since!

In Agra we checked into our hotel in the late afternoon and were told that there was a party beginning in a couple of hours' time. What with all the anxiety about the delayed fights and so on we decided to take a nap for an hour or so—of course, when we woke up several hours later the party was well and truly over. However our visit to the Taj Mahal the next morning made up many times over for any disappointment. It was the experience of a lifetime. The rich decoration of the structure using semi-precious stones and marble was absolute stunning.

****

After returning to Calcutta from Puri we had busied ourselves doing the groundwork for our return to the UK, but there was one major hurdle. At that time—in the mid-1960s—there was an extensive so-called brain drain from India, in which Indian professionals and specialists were emigrating to Western countries for better opportunities and a better life, particularly if they possessed foreign qualifications. To try to put a brake on this, the Indian government was inventing various ways of discouraging emigration.

Nevertheless, I decided to resign from the company. My employment contract stated that I must give six months' notice; however, my superiors asked me not to make my departure general knowledge just yet, for fear that my subordinates would not comply with my directives during the notice period.

As I had expected, my Indian family were extremely unhappy about our decision. In order to avoid the government's embargo on emigration, I told the authorities that I was visiting the UK temporarily to attend my sister-in-law's wedding. In addition to the embargo, transfer of currency for me was limited to £3.50, although Brigid and the children obtained a slightly more favourable permit for transferring currency because they would be travelling on her British passport. I also had to produce a letter from my father-in-law stating that he would be responsible for any expenses incurred by us while we were his guests in the UK.

We sold whatever household goods my Indian family didn't want, and expat friends in India converted our Indian currency into pounds sterling and deposited the amount into our UK bank account. The Indian taxation office had to certify that I didn't owe them any money to allow me to get an exit permit; and once again, our sea passage to the UK had to be booked well in advance of our departure in March 1965.

****

Voyage to Britain

It was a very sad occasion when we had to farewell our family members and leave our Calcutta house for the train to Bombay, this time with five children instead of four. Due to the foreign currency limitations, our spending on board ship had to be limited.

The *Orcades* docked for a few hours in Lisbon and Barcelona, among other ports, and passengers were able to go sightseeing. At that time there was ongoing government-level conflict between India and Portugal over the disputed territory of Goa, which India had annexed from Portugal in 1961. As a result, since I had an Indian passport I could not disembark in Lisbon, so Brigid did a sightseeing tour while I looked after the children. I went ashore in Barcelona while she did the babysitting, and I visited the four centuries-old Cordorníu winery, the home of Spanish champagne, or *cava*, which we very much enjoyed later.

PART 4
Back in Britain

The rest of our journey was relatively uneventful and we were delighted to be back in England after nearly two and half years.

My in-laws most kindly took us into their home, a small cottage in the extensive, beautiful grounds of Wynstones, a Rudolf Steiner (or Waldorf) School in Brookthorpe, Gloucestershire, until we found our feet. It was difficult to fit us all in—seven of us, and five already living there—but there was also a small place in the grounds to cater for overflow accommodation, so some of us could sleep there. The cottage belonged to the school. Brigid's father was one of the founder-teachers, and Brigid had herself attended the school. There were several other cottages in the school grounds and a hostel for the schoolchildren, all of which provided a community setting in line with Rudolf Steiner's principles.

I had two immediate tasks. One was, of course, to find a job so that we would be self-sufficient before our savings ran out. Secondly, my Indian passport was about to expire, so I would have to return to India within three weeks unless I could get a British passport before then. Our solicitor set these particular wheels in motion; meanwhile the children were invited to attend the school, and were very happy indeed to be going to their mother's old school. Their fees were temporarily waived until I started earning money.

It wasn't easy for a foreigner with very specialised qualifications and expertise to get a job in England at that time. I began by contacting Dr Bob Westwater, who was fairly high up in the ICI Group. He advised me to contact Joe Abrahams, managing director of a small company called Rock Fall, which was at that time in the process of expansion. Joe Abrahams was a very busy man but, when I did eventually manage to speak to him, he invited me to travel to Glasgow for an interview, timed to take place while he was in transit between flights to and from one place to another.

The interview was conducted in the bedroom of the hotel where I was staying. It went well, and he offered me a position as Explosives Engineer. He explained that the company's operations involved a wide range of drilling and blasting contracts, and they had recently embarked on a specialised project involving underwater rock excavation by blasting, and I was earmarked for that. Abrahams assured me that financial assistance would be provided to buy a house for my family, and that, since the work would not be tied to any particular place, there was no requirement to live in Scotland if we chose not to. The children were by now happily settled at Wynstones School, so we decided to look for a house near Brookthorpe.

My application for British citizenship had meanwhile been successful on the grounds of the long period of my stay and work there prior to leaving for India. This, together with the job offer, was a great relief.

I was told to spend a few days at Rock Fall's head office to familiarise myself with the company's structure and personnel before becoming actively involved in their projects.

When I arrived in the company's head office in Glasgow I got a shock. It was located in the Shawbridge district, at that time a run-down part of Glasgow, and the building was not much more than a glorified work shed. What a contrast to my previous office in Calcutta, which was in a modern air-conditioned building! However, beggars can't be choosers, and I had to accept the situation. The staff turned out to be very friendly, and they brought me up to date on the company's structure and the different divisions such as Civil Engineering, Marine, Demolition, Exploration and the like, depending on the type of contracting work, so that I soon had a good grasp of all the company's activities.

****

Job with Rock Fall

My first assignment was in the Marine Division, which required me to travel to an underwater drilling and blasting site in Middlesborough, North Yorkshire. The operation was carried out there from a barge carrying four large overburden drilling rigs. The barges had between one and six booms, depending on the size of the contract. (Photo 20)

**Photo 20: A typical 6-boom underwater drilling barge
(Courtesy: Rock Fall Co)**

The object of this operation was to deepen the harbour by blasting and removing the blasted rock, to enable larger capacity vessels to dock for loading and unloading on the River Tees. This was carried out over sixteen hours each day, worked in two eight-hour shifts. There was a portable project office on shore, and communication with the foreman on the barge was by walkie-talkie. The foreman logged the progress of the work in a small office on the barge. A powerful launch was used to ferry personnel and materials between the shore and the barge, and also to tow the barge when necessary, and for positioning the barge's anchors. (Photo 21)

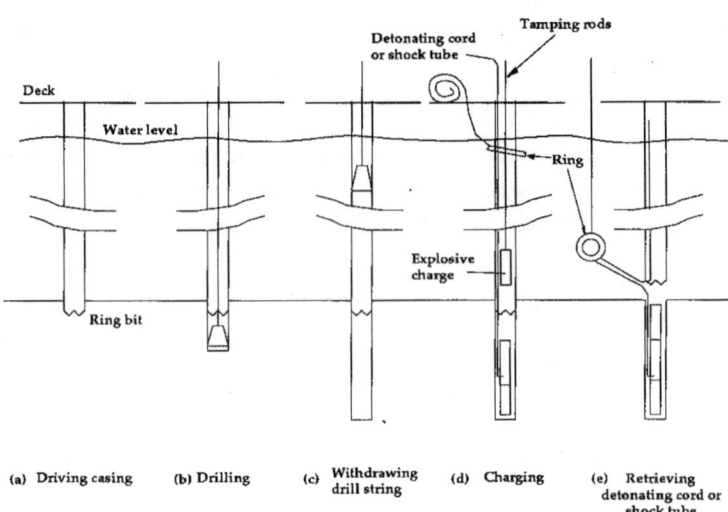

(a) Driving casing　　(b) Drilling　　(c) Withdrawing drill string　　(d) Charging　　(e) Retrieving detonating cord or shock tube

Photo21: Sequence of operation for loading explosives

The figure shows a typical operational sequence. The first step was to drill through the mud and water on the harbour floor using hollow drill rods. This removed the mud, and penetrated a few millimetres into the rock, effectively placing and stabilising a hollow tube from the water surface to the bedrock. Solid drill rods were then pushed through the tube so that the blast hole could be drilled in the underlying rock. The required number of holes—in this case, four holes—were drilled in this way, and each was charged with the calculated amount of explosives connected by a detonating cord; blasting was delayed until all holes were charged with explosives. The four detonating cords were then joined to a trunk detonating cord. The launch then towed the barge to a location a safe distance away, with the trunk cord being fed out the all the while. A warning siren was then sounded shortly prior to the explosion.

One of my tasks from the shore was to ensure the safety of the area was for blasting, and to warn all parties in the near vicinity prior to the explosion. After the blast, the barge was towed to a new location where more of the harbour bed needed to be broken up by blasting. Its new required position was accurately determined relative to precisely located onshore markers by using a sextant, and the drill/blast sequence was repeated.

****

The Institute of Explosives Engineers was formed a few years before I joined Rock Fall. One of our senior staff was a member of the Institute council. I was asked to replace him, which meant my attending the monthly council meetings. At one meeting I proposed that the institute should run a shotfiring training course to help establish a high standard of blasting operations. This was accepted, and I was tasked with organising the inaugural blasting course on behalf of the Institute. To do this I gathered a number of experienced speakers covering a range of blasting operations. The course was well attended, and so successful that follow-up courses were organised by the Institute in following years.

****

For my submarine blasting contract I was mainly occupied with onshore administrative work. One day, however, I decided to gain a more precise knowledge of operations on the barge. The barge supervisor, Jimmy Doyle, sent the launch to ferry me over to the barge. When I arrived, I thought my eardrums would rupture from the noise from the simultaneous working of four drills, so the supervisor and I took shelter in the small cabin where there was a little less noise. Pointing through a window, he began explaining the operation.

Doyle suddenly exited the cabin mid-sentence, and began speaking rapidly to one of the drillers. I was a little surprised by this odd behaviour, but thought there must be a reason; shortly afterwards he returned to the cabin and apologised for his sudden disappearance. It seemed that he had realised from the drilling noise that one of the drill bits was about to get jammed in the rock. I asked him how he could separate one drilling noise from another. He explained that the safe drilling procedure was to drill the hole to a certain depth and then to flush the debris from the hole before continuing to drill. Occasionally one of the drillers wanted to finish his hole before the others and would try drilling deeper before flushing. This procedure risked jamming the drill. The sounds from all the drills were like an orchestra, Jimmy Doyle said. If one of the drills were being pushed too far its sound would be quite noticeably different to an experienced ear. If he allowed this to continue, a great deal of time and effort would be wasted in releasing the drill bit, and the other drillers would have to sit idle, so he had to step in quickly to avert the problem.

This is the sort of knowledge—distinguishing between an 'out of tune sound' and a smooth running of the drills—that cannot be gained from books or classroom lectures. Doyle also showed me many other practical aspects of the barge operation; later I discovered that he had had little formal education due to his parents' financial position, and that he had had to start working at the age of fifteen. In my estimation he was a very clever man, and I have always admired him for passing on the knowledge he had acquired through experience.

Gradually I became fairly expert at this type of marine excavation work, but the downside of this job was living away from home and staying in hotels most of the time. My routine work schedule was such that I would start on the afternoon shift on a Monday, then work twelve days straight. At the end of the morning shift on the Friday twelve days later, I would drive home to Gloucestershire for three nights with Brigid and the children. These breaks were a great joy to me.

****

In the middle of 1965 we found a semi-detached house for sale in the Gloucestershire countryside. Called "Donard", it was in Pitchcombe, about three miles from Brookthorpe. My

Photo 22: "Donard" – Our first home purchase

company assisted with an interest-free loan towards the deposit for this, the first home we had ever owned. Photo 22

We had little furniture, but our first priority was to install a central heating system. This caused problems with the cinder-block walls. The unaccustomed warmth dried the interior plaster, which cracked and started to fall off causing the cinder aggregate spilt out. The original material had to be removed and the walls re-plastered—expensive and extremely dusty—so back we went to the in-laws for a week or so while the work was being done.

****

One marine project involved blasting rock only one metre above the crown of a service tunnel under the River Tees used for transporting materials between two branches of a large industrial complex located at either end of the tunnel. Any damage to the tunnel caused by vibrations from the blasting would bring both factories to a standstill, at heavy cost.

This delicate operation required strict compliance with ground vibration specifications. The work took about two weeks, during which time I had sole responsibility for deciding the amount of explosives to be used. I also monitored the blast-induced ground vibrations inside the tunnel, as did two independent agencies employed by the client (ICI).

We maintained vibration levels within the safety margin, and no cracks or other damage occurred in the tunnel; at the end of the contract, the client sent a very appreciative letter.

Another contract involved a strict deadline, with a heavy penalty if this deadline were not met. All port facilities were suspended for the duration of our blasting operations and the affected vessels had to wait outside the river estuary. In order to complete the job within the specified period, a continuous work schedule of two 12-hour shifts was placed. I was the site agent for the job and I spent something like 18 hours a day making sure that there were no hiccups. Work progressed well and I expected that we would meet the completion date.

One evening after dinner I was in the shore cabin doing some paperwork when I got a call from the supervisor of the barge. He announced that the workmen had found out about the strict time frame for finishing the contract and, taking advantage of this, had stopped work and were demanding a pay rise. This was a shock to me, and at first I was at a loss as to how to proceed. I asked the supervisor to play for time by starting to negotiate. I realised that I could not possibly agree to a pay rise as that would have jeopardised the company's pay structure for all other sites.

It was almost midnight at the time so I could not phone the MD for advice. I flicked through the contract document for the job and came across a clause to the effect that, in the event of an industrial dispute, "... there would be no demurrage claim for not finishing the work on time provided we made every effort to continue the work". This gave me the grounds I needed for taking appropriate action—I immediately sent a message to the barge supervisor saying that if the workmen were unhappy with the current pay scale, they could leave. I saw the men come off the launch and walk pass my office, swearing at me in extraordinarily colourful language.

I then arranged for all officials, including myself, to take over the working of the rigs as best we could the next morning, in order to honour the terms of the contract. On that basis, work proceeded continuously, although progress was very slow. At the end of the day some of the men returned to work. All but two eventually returned and, once we had put that episode behind us, the progress of the project was satisfactory.

On another occasion I was working on a large contract where the site agent, Eric O'Connor, had served the company for a long time. All along my policy was to reward the men by allowing them to finish early if the day's work had been completed, as long as the quality of their performance was not compromised. O'Connor did not approve of this policy, nor of some other minor actions I had taken. Relations between us soured to the point that I could no longer work at the site. I communicated my feelings to the MD who transferred me to a contract for the Civil Engineering Division in Richmond, Yorkshire.

This was a road-making operation in very hilly country and was a relatively straightforward job. Shortly afterwards I heard news of a barge explosion in our Marine Division, causing the death of a driller, at the Middlesborough site I had just left; a few days later I was requested to rejoin the Marine Division. This small break from the Marine Division worked wonders for me. I found that all the senior staff were more amicable towards me, and from then on I was quite happy to work at that site.

6

School Teacher-cum-Consultant

During one of my weekend breaks at home, Brigid told me that the senior mathematics teacher at Wynstones School was leaving for an overseas appointment, and someone at the school wished to discuss it with me.

I then had a visit from Paddy Lewers, a senior member of staff at the school, who explained that the loss of this member of staff was sudden and unexpected, and that there wasn't anyone on the immediate horizon to fill the gap before the next school term began in a matter of weeks. Would I consider taking the post at least for one year until a suitable person was found? He spelt out the terms and conditions; I promised to let him know my decision before the weekend was over.

I discussed the pros and cons of this new direction in Brigid. The biggest problem would be the drop in salary, which was only about half of what I was currently getting. The positives were that my base would be at home, and this meant that I could be more time with my family; there were three long holidays in a year, and staff members did not have to pay their children's school fees. In reality, the first point carried more weight than the others; moreover, there was some chance for me to do outside work to get extra income. Having considered all these factors, I decided to accept.

When I told my boss what I intended to do he seemed disappointed, but suggested that I might like to continue my association with the company as a consultant. By that time I had chalked up considerable experience of their kind of work and I might be useful to them from time to time if I could make myself available—during school holidays, for instance. I welcomed this idea and looked forward to continuing my involvement with industry. I resigned from Rock Fall at the end of August 1966 to become a teacher at Wynstones School.

My knowledge of the school mathematics curriculum after more than ten years was very rusty, so I had to re-learn a lot of it—hard work for a brain going on for forty years of age! I remembered hating geometry at school and relying on my stronger arithmetic and algebra to get me through Mathematics exams; now I was going to be teaching the very subject I had neglected and despised in my teens. However, I found that I didn't have any difficulty with the geometry— it was only a school-level subject, after all, not rocket science— which just goes to show that one can grasp a topic quite easily at a mature age.

I would be teaching Mathematics to Upper School Classes 9 to 12 (ages 15–18) and preparing some of them for university entrance. I was also to be a "class sponsor" of the current Class 9. Class sponsors were responsible for the performance of all the students in their class, and assisted anyone having difficulties or problems. Sponsors stayed with their class for the four years from Classes 9 to 12, which helped to forge a lasting relationship between each sponsor and their pupils.

A feature of the Waldorf Rudolf Steiner curriculum, based on Steiner's anthroposophy,[2] was to provide the "main lesson" material to each class in the first two hours every

2. A philosophy developed by Rudolf Steiner that, among other things, "... aims to develop faculties of perceptive imagination, inspiration and intuition through cultivating a form of thinking independent of sensory experience, and to present the results thus derived in a manner subject to rational verification" (see e.g. *The Essential Steiner* (1984) by Robert McDermott).

morning. This covered a particular theme for two or more weeks, during which time it followed fundamental principles that students grasped easily. It was often the practice to introduce artistic elements into the theme content. Students prepared a book for every main lesson theme from notes they had taken in the main lesson, and which they might have followed up by their own private study. Main lesson periods were such that they required the teacher to devote a considerable amount of time to lesson preparation.

As our own children grew up and reached the upper school, I was their mathematics teacher. This situation was not always easy for either the children or for me; sometimes, for example, I purposely did not give my children's achievements the praise they deserved as a demonstration to the class that I was impartial and that there was no favouritism. Most of the time, however, our relationships at home was loving ones.

In the meantime, Brigid and I were very aware of the lack of space in our home, and we were looking for a bigger place that would still be affordable. An opportunity came when Jenny De Havas, whose children were at Wynstones, decided to sell "The Limes", her eight-bedroom Georgian house in Brookthorpe. As well as being spacious, it was very close to the school and we would no longer have to travel three miles each way. Since she had a connection with the school, her asking price was very reasonable although naturally we knew that there would be still be a big gap between the selling price of our home and the purchase price of such a big house. However, if we were to provide accommodation for some school boarders, we should be able to cope with the extra outlay; we went ahead and bought "The Limes", to the great relief of the children especially, as they could now all have their own bedroom. (Photo 23)

Eventually we had up to four students boarding with us. Having five children of our own as well as being guardians to four boarders was something of a challenge at times: our

Photo 23: Our next abode is a Georgian house, The Limes

home became almost like a mini-hostel. We introduced a scheme where every child had to do a small household chore on Saturdays; in return, they could choose their favourite meal at weekends. We also acquired a dog around that time, a Labrador-Collie cross called Sheba—a source of good fun, especially for the children.

Later I joined the local rowing club and I arranged for them to let children in the Upper School use their rowing facilities. I took a group in a minibus once a week to row in the local canal for about two hours. This was a great success, offering as it did another sporting facility for the older boys and girls.

****

As arranged, I worked as a consultant for Rock Fall in most school holidays on short-term contract assignments. Some of these projects were very exciting, and I have included a description of them here.

1. One project involved finding suitable rock for the Jubail Harbour Project in Saudi Arabia. This was not only to construct a harbour but also to build a long

breakwater to protect the harbour from storms. Our expatriate team consisted of a Welsh geologist and me. Head Office had already arranged for equipment and manpower to be available in Dhahran. We were to travel from the UK to Abu Dhabi, where we would collect our Saudi visas for Dhahran. Since this was a Saudi-funded project, the government insisted that the necessary rock be excavated in Saudi Arabia itself.

The geological map of the kingdom showed that un-weathered bedrock was overlain by very deep sand except in one area near Dhahran, where there was a slim possibility of finding bedrock at a shallower depth. However, after about ten days of drilling and geological examination of the rock chips from the drill-holes, it was clear that the rock would not be suitable for harbour construction work and we had no option but to abandon the site with no net result. Work then moved to the United Arab Emirates.

There it was found out that nearest competent rock deposit was to be found in Ras al Khaimah. After some scouting expeditions, we found an area close to a sea inlet where suitable rock could be quarried. As the surveyors were establishing the proposed quarry boundary, we were startled by rocks being thrown at us. Luckily, someone in our group spoke the local language and shouted out, asking who was throwing the rocks. It turned out that they were nomadic Bedouins who were very suspicious of strangers and wanted to drive us out of their territory. When we explained what we were doing there was really for their benefit, they left us alone.

It took some time to mobilise our equipment and labour force. In due course we started the drilling operation to sample the bedrock, which we found to be competent and meant that our mission had been successful and we were able to return home. Eventually a full-scale quarry operation was established at the site to supply rock for the Jubail Harbour project, for sea transport via the Persian Gulf to the project site.

I was staying at a Western-style hotel that boasted the only casino in an Arab country at that time. Late each afternoon a number of small private planes would land in the grounds of the hotel and their passengers headed straight for the casino. One day, one of the Managers from Head Office, Peter Flynn, arrived to inspect the progress of the work. Over a glass of beer he asked me whether I had visited the casino. I told him I couldn't afford to treat myself to that kind of thing. He gave me £150 (as a special allowance) to try my luck at roulette; so, to have a one-off lifetime experience, I joined a group at a roulette table and promptly lost the whole lot in less than ten minutes! I then just watched what other people were doing. The room was mainly filled with Arabs recently arrived in their private jets from all over the Middle East, wearing the traditional burnous, their pockets jammed with dirham notes, the famous petro-dollars .

2. Another project was to become the scene of my greatest professional folly.

I was in charge of a project to widen the fishing harbour in Lochinver, north-western Scotland. It involved levelling a stone outcrop adjacent to the harbour jetty by blasting small volumes of rock at a time to safeguard existing structures. Shot holes were to be drilled into the rock and loaded with explosives, then the shot-hole area was to be covered with a protective steel net anchored to the rock.

In the harbour, the daily routine was for the catch of the day to be unloaded from the trawlers in crates onto the jetty. The fish were then auctioned and loaded into trucks to be dispatched to the markets. Our practice was to suspend our activities for as long as the loaded fish crates remained on the platform. One Saturday morning the trawlers unloaded their fish crates onto the wharf and the fish were auctioned. Unusually, however, everyone left the jetty without shifting a single crate.

I assumed that there might be some dispute that had to be settled before the crates could be dispatched, but more than half an hour passed and there was no sign of the fishermen or their customers. I could hear a faint noise coming from a pub about four hundred metres away, so I sent one of my men to the pub to see if he could find out what was happening, and he reported that all the men were drinking and having such a good time that they were in no mood to do any business at that point. In the meantime we had been idle for nearly two hours. I was in a quandary— how was I to justify this lack of work progress in the logbook? On the other hand, we couldn't just leave, with the holes charged with explosives, and go home—so I decided not to waste any more un- productive time, and let the shot off.

As usual, the warning siren was set off, and I could see the whole crowd at the pub peering at us through the doors and windows. We took shelter and let the shot go off. Oh, hell—what a sight! Part of the safety net must have become loose, and a massive amount of small rock debris and dust shot up and landed on the jetty on the fish crates. I was so shocked that I wanted to take the coward's way out and run! But I wasn't going to do that, so I prepared myself to hear the verdict. By this time all the merrymakers were examining the damage and one of the fishermen's leaders declared that the blast damage to the fish amounted to nearly £8,000—a lot of money in the 1960s. It was covered by company insurance, but nevertheless I was not popular with my bosses. From this incident I learnt a valuable lesson: NEVER LOSE PATIENCE!

3. Around that time, I became involved in the extraordinary case of the wet carpet.

In the strict security period that operated during the Troubles in Ireland in the late 1960s, my company was contracted to deepen the area in front of the jetty in Howth, just north of Dublin, to let larger fishing trawlers dock in the harbour. This, again, was a task that required drilling and blasting the bedrock, with the broken rock then dredged for disposal elsewhere.

The sectarian violence of that time meant that there were severe restrictions on the use of explosives, and the explosives stock supplied by the manufacturers had to be stored at an extremely secure army depot. Every morning our van collected explosives from the depot and was then escorted by two armed police vehicles, one in front and the other behind the van. We had to estimate the full amount of explosives to be used that day, and all of it had to be transferred from the van to the drilling barge. An armed police guard was posted at the site throughout the shift to ensure that no explosives could be taken off the barge.

Any surplus explosives had to be destroyed at the end of the day—obviously, for the security of the depot, they would not accept any returns. The work procedure was routine, except that we could only work one shift each day. The Howth harbour area was home to wealthy retirees whom we had to make sure were not upset. Before starting the blasting work, we photographed every house near the jetty and took note of any defects, to safeguard the company from false claims for blast damage.

One day I was in the office when I received a telephone call from a lady who demanded compensation for damage to the carpet in her lounge room caused by our blasting. This was puzzling, and I must admit to being just a little incredulous, but nevertheless I arranged a suitable time to inspect the damage.

When I arrived at her house, she took me to the lounge room and, near the bay window, showed me a damp patch on the carpet. I thought that we might have damaged the outside structure of the house, letting rainwater seep through to the carpet. I was about to go outside to check this possibility, when she stopped me and said that the damp was due to her dog urinating: she explained that as soon as our warning siren sounded, her dog used to charge straight into the lounge room, look out the window, and tensely wait for the blast—bang! And it urinated on the carpet. To verify this extraordinary story, I went to the house shortly before a blast, and, sure enough, witnessed the whole episode, just as she told it. After several such performances, the carpet had deteriorated visibly. Naturally I had to admit the company's liability for the damage. I made a deal with her: if she agreed to wait until the blasting work was complete, she could buy any carpet of her choice, and the company would reimburse the cost.

4. At another time, still in Ireland, my colleagues witnessed the following sad story.

One of our engineers, from a Protestant family, had been going out for a while with a girl from a Catholic family in Howth. They decided to marry, and her parents arranged the wedding. I had been invited but couldn't attend that day due to a shortage of supervisory staff on the project.

To my surprise, my colleagues who went to the ceremony came back early, and they explained why. It seems that the night before the wedding, the local Catholic priest had instructed the bridegroom on the responsibilities and duties that he would be pledging himself to undertake when marrying into the Catholic faith. For instance, after the wedding the couple's children had to be raised as Catholics, go to Catholic schools ... and so on. He listened to all these conditions, and accepted—outwardly, at least.

The next day, suitably attired, he went to the church and took his place before the altar as they had rehearsed ... Suddenly the organ began to play as the radiant bride arrived; the congregation stood as she proceeded slowly up the aisle on the arm of her father, with all her bridesmaids in attendance on this wonderful day—her day of days—and joined her husband-to-be at the altar. When the priest turned to face the bridegroom and asked him if he would take this woman to be his lawful wedded wife, without a word and to everyone's shocked disbelief, he suddenly turned on his heel and bolted out of the church. Before he left Howth for good he wrote a short note for everyone, explaining that he had had a change of heart about converting to Catholicism and could not accept the conditions imposed by the Catholic church; he had changed his mind at the last minute in order to avoid a possible rift in the future. This was as distressing for him as for the wedding guests—and it seems that no-one had the courage to enquire about the bride or her family.

****

A Taste of freshly caught wild Salmon

During the school holidays (mostly at Easter) when there was no call for a consultancy assignment, we used to explore various parts of Britain, often by bicycle. Since our finances were rather tight, we decided to use Youth Hostel Association hostels instead of hotels, so we joined the Association as Life Members. This membership allows us to use YHA accommodation throughout the world to this day. The YHA had recently modified its rules at that time, so that members were no longer required to travel as walkers or cyclists but could use motorised transport as well, although members were still only permitted a maximum stay of three nights at any one hostel.

Our method of travel was this. We used to put the cycles on top of the car and then drive to a youth hostel, stay the three nights allowed, and explore the area by day before moving on again. In this way we gained an intimate knowledge of many parts of the country.

In one occasion we were exploring Scottish Highlands, although this time without bicycles. We stayed in a small hostel near Ullapool in the far north. One morning I went down to the jetty and saw that the local fishermen had brought in their catch of beautiful gleaming salmon, freshly caught. I asked if I could buy some, but the fisherman'said the whole catch was already spoken for and that he couldn't sell me anything. In fact the boat that was coming to collect the catch could be seen in the distance. The thought of enjoying such fresh and gleaming river salmon emboldened me and I redoubled my pleas, saying it was the chance of a lifetime and how much I loved fish of any kind but fresh salmon was something really special. While I was arguing my case and pleading with the fisherman to let me have at least a couple of steaks, the man's wife came to my aid, saying: "Let the lad have some. We never get the chance to eat salmon ourselves either so this'll be a chance for us too. Let him have a couple of steaks, we'll have the rest". And the fisherman did!

That night we cooked wild salmon cutlets in the youth hostel kitchen to the envy and amazement of all the other hostellers. It was a meal to remember, and I still do.

****

Occasionally I was called upon even during the school term to do consultancy work lasting just a few days. If I could, without too much disturbance to the students' progress, I accepted it.

On one occasion, towards the end of June when I was extremely busy with students' progress reports, and with their exams looming I was asked to do an urgent job in the Hebrides in northern Scotland; regretfully I had to decline. Unfortunately this didn't go down well with the company, and that was the end of my consultancy days with them—I was never asked again.

7

Another Turn in the Road

One after another our own children left home for tertiary education or to follow a career path. Anita, our eldest daughter, became a qualified architect; our eldest son, Prasanta, took up mechanical engineering; and our middle daughter (who had by now changed her name from Runa to Jane) had begun a degree course in German and Arabic. The two younger ones were still at school.

One day in April 1978 I was busy preparing myself for the summer term's school curriculum when I received a call from Rock Fall to tell me that their Technical Service Manager, Peter Harrison, had suddenly decided to resign, and they were offering the post to me, ahead of advertising it openly. This news was a complete surprise to me; I promised to call them back within forty-eight hours. The company by this time had expanded very rapidly, and my responsibility would be the same as being a consultant on a retainer, which meant that I would be a full-time trouble-shooter, ironing out problems.

This raised some very difficult points for consideration. On the one hand I was reasonably settled in a house very close to the school, and Brigid was also teaching nearby. Although three of our children had left home, two of them were still at school: our daughter, Shakuntala (nicknamed

Shuki) was in her final year at Wynstones, and Rahul, our youngest son, was at Marling, a grammar school in Stroud. Why would one want to upset this equilibrium?

On the other hand, if I missed this opportunity, it would be very difficult to later get back into the profession for which I was qualified. There were other issues as well. To accept the position would mean uprooting and moving to Scotland where Rock Fall was based, and that had to be considered. There is a big difference between Scottish and English school curriculums: would it harm our children's progress if we were to plonk them into Scottish schools? Since the position was vacant the company wanted to fill it as soon as possible but it would be a betrayal to the school, in particular to the students, if I were to join the company in mid-session.

On the positive side for rejoining the company, I had by now been teaching at Wynstones for nearly twelve years, and most of the teachers who were my contemporaries had either retired or moved on; there was a large group of new young teachers who had rather radical views, and I was uncertain whether I would be able to tune into these new objectives; furthermore, if I declined the offer my toehold in the industry might be lost forever. It happened that I had been the class sponsor for a group of students who would reach the end of their schooling in about three months' time, so a decision to continue at the school would mean taking on a new class for a further four years.

Considering both the pros and the cons of this new move, I decided to accept the offer on four conditions. The first was that my commitments at the school meant that I could not start before the end of July. We had finally decided that the children's schooling should be completed in England, and so the other three conditions were all about reimbursements: payment of Shuki's final year school fees at Wynstones, my hotel expenses and the cost of visiting my family at weekends for at least six months while I found a house for us all, and lastly, relocation expenses from England to Scotland. Rock

Fall agreed to all these conditions, and we also agreed that I would take up my new position in August.

****

At the school's College Meeting (its working committee) it was with great trepidation that I announced my intention to leave. It came as a shock to everybody there. In the silence that followed I justified my decision by saying that the time had come for me to move on. It was not easy for me to sever my connection with the school after twelve years' service.

The new position entailed determining if there were any blasting-related problems at any of the company's work sites, and required a fair amount of travelling. Since it was unlikely that my family would be able to move to Scotland for the next three years or so, we would have to have homes in both England and Scotland so that we could visit each other; so whenever I was in my office in Barrhead (a few miles from Glasgow) I looked at houses for sale in the area.

Eventually we found Grougarbank House, a two-storey duplex stone building on the bank of River Ayr, on about

Photo 24: "Grougarbank House"-Our second house in Scotland

two acres of land (Photo 24) in Hurlford, about twelve miles from head office. In the garden were roses and a large area of beautiful rhododendrons.

An older couple owned the top floor and had been living there for many years, and the ground floor was owned by an elderly man who was now permanently in a nursing home, and it was now for sale, fully furnished—ideal for me.

The property had not been lived in or heated for such a long time that the damp was very evident, so I installed wood-burning stoves in every room and had the entire ground floor area damp-proofed using the electro-osmosis method. a wood-fired cooking range was also installed in the kitchen, but it didn't prove to be very successful.

Since I was living on my own most of the time, I brought our dog, Sheba, from England to keep me company. Because I had to leave her behind whenever I had to go away on company business, I first of all tried leaving her in a kennel, but she always looked very miserable when I retrieved her, so a colleague and his wife, who were also very good friends, offered to take care of her during my absences from home.

After I moved to Scotland, The Limes was now too big for Brigid and the two children, Shuki and Rahul, so in due course we sold it. Brigid, who was still teaching at Stroud College, found an almost derelict stone building called Elm Tree Cottage next door to the college, for sale by tender—that is, where prospective buyers lodge a written offer by a certain date, to be opened only after the deadline. The highest bidder is successful and is able to buy the property. We realised that the property would need a huge amount of work to be done to it, and that the restoration would require a lot of time and money. To decide our offer, we estimated the value of the cottage as it stood and the cost of restoring it, then wrote down figures at £100 intervals on ten pieces of paper, picked one at random and tendered the figure written on it. This method worked in our favour, extremely luckily, and we became the successful bidders!

Brigid found a builder for the rebuilding work, and Anita, our architect daughter, took on the redesign of the project. While the building work was going on, Brigid lived in a caravan parked in a friend's garden next door to her mother's house, and Shuki and Rahul slept at their grandmother's house.

It took almost a year to completely restore the cottage. (Photo 25)

Photo 25: Elm Tree Cottage after finishing the renovation

As well as renovating the three existing bedrooms, a large attic room was converted into a fourth bedroom and an en suite bathroom was added to the master bedroom. A gas-fired Rayburn provided cooking facilities, central heating and running hot water. The cottage was comfortable for the four of us, and it could temporarily accommodate our whole family; a big bonus for Brigid, of course, was that she could walk to work at the college in five minutes.

****

When I rejoined the company full-time, it was housed in its new purpose-built modern building, and I began to realise that the structure of the company had now become top-heavy. Then at the beginning of the 1980s came the economic recession in the UK, together with rising unemployment. Our company was also failing to procure new contracts, presumably as our tender prices were too high and were therefore not competitive; in my opinion the reason that our tender prices were excessive was because of the company's high overheads.

Eventually the threat of redundancy gradually became a reality, and the office was reduced to a skeleton staff. There were hardly any jobs in which our company was involved. Then came a tentative opportunity to be involved in a big project in offshore ocean floor manganese nodule mining. With this in mind, I was invited to attend discussions at our group headquarters in Holland, during which it transpired that if that project did materialise, staff would have to work six months offshore followed by six months' leave. I was not too thrilled by the prospect of such working conditions, and started looking around for alternative job opportunities. Although I was one of the last members of staff to be made redundant, those last few weeks were rather distressing because of the lack of any productive work or sense of direction.

8

A New Opportunity in Australia

An advertisement in my professional journal announced a vacancy for a lecturer's position in mining engineering at the University of New South Wales (UNSW) in Sydney, Australia. I responded by writing a half-page letter expressing my interest in the post. Shortly thereafter I received a three-page letter from Professor Frank Roxborough, Head of the School of Mining Engineering, giving details of the position and saying that he hoped that I was serious in my inquiries, as they really wanted me. I had known Frank since the Newcastle days nearly twenty years before—as I have already said, we celebrated our PhD graduation at Durham together on the same day.

I was asked to complete an application form together with a résumé of my work experience. Frank advised me to request a tenured position rather than a three-year fixed-term appointment. He explained that if I took a fixed-term appointment, I would be given relocation expenses both ways but after three years the position would be re-advertised and, if I wanted to stay, I would be just one of the applicants for the job, which carried with it the possibility that I might not be successful.

After discussing it with Brigid, I finally decided to apply for the fixed-term position nevertheless, as it seemed such a gamble otherwise. If I didn't like the job, or if we didn't like being so far from our family, at least we would not have to pay all our own relocation expenses back to the UK with a fixed term contract. In due course the letter of offer arrived from the university, and we also received our temporary working visas from the Australian Government. Brigid had decided to continue teaching at the college next door for the time being to take care of Rahul, who still had another two years to go before finishing school at Marling.

I then started the serious organisation involved in emigrating to Australia, timed to start my lecturing duties in the second semester starting in July of that year, 1981. Brigid would follow me later on. My flight was in a month's time and I had many loose ends to tie up as well as bidding farewell to friends and relatives.

After qualifying as a mechanical engineer, our eldest son, Prasanta, could not find a suitable job, and was filling in time by doing temporary work. Whilst he was studying in Leicester he had met a Japanese girl, Fumiko, who was an English Language student. She told him that there was a huge demand for English teachers in Japan, so he decided to go there and teach English.

Photo 26: Sen Family in 1981

One day Prasanta brought Fumiko home and introduced her to us. She seemed to be a sweet girl who would help him in his desire to live in Japan. Shortly before I left for Australia she took a group photograph of the Sen family together with our dog, Sheba. This, from 1981, is the last photograph taken in which the whole Sen family was present. (Photo 26)

I then arranged with a shipping agent to send our car, a Citroën CX, to Sydney, filled with various belongings. Soon after that, Prasanta left England for Tokyo.

PART 5

In Australia

9

Arriving in Sydney

I touched down in Sydney on a cold morning on 12th July, 1981. One of my new academic colleagues, Dr Amal Bhattacharyya, met me at the airport and took me to Basser College, one of several student residential colleges on the university campus. Prior to my arrival in Sydney, Professor Frank Roxborough had arranged with the college authorities for me to have a tutor's flat in the college. My duty there would be to assist and to take care of the students as and when necessary. This self-contained, one-and-a-half bedroom flat was ideal for me. Food was provided in the college canteen so I rarely cooked, giving me a chance to spend more time on preparing my lecture notes.

Besides Basser College, two other Colleges, Goldstein and Baxter, were part of the Kensington College complex. At the head of each college was its Warden, and the overall head of Kensington Colleges was the Master, Ken Bromham. A naval officer, Commander Ted Shimmins, who was at the university on a temporary assignment, was the Warden of Basser College when I arrived there.

At the end of my first semester I returned to the UK to be with the family for the Christmas holidays. En route to the UK I stopped over in India to spend a few days with my Calcutta family. It was a great joy to see my mother and the

**Photo 27: Gour and Brigid with Calcutta family in the courtyard
of 3/3A Shanagar Road**

other members of family after a gap of 16 years. (Photo 27)

Brigid joined me in Australia during the northern hemisphere summer holiday (July–August) from Stroud College, and we took a short holiday to Fiji. While we were there I visited the Emperor Gold Mine (now Vatukoula Gold Mine) and presented a short seminar for the staff there. This "to and fro" style of living continued for another year.

At this point, the following year, Rahul flew to Sydney on a one-way ticket to spend his gap year between school and university. I had arranged a mining job for him in a gold mine in Northern Territory, once he had spent a couple of weeks in Sydney, with the idea that his good wages for three months at the mine would take care of his return fare. However, Rahul found a job in the hospitality sector in Sydney and quickly became so involved with life there that he decided to stay on in Sydney rather than go mining. In the meantime Brigid had resigned from Stroud College, and she finally joined me in Sydney in July 1983.

A new warden, Josh Owen, was appointed to Basser College in 1983. We found working with him difficult, so decided to buy a one-bedroom apartment close to the university and move in there. We soon found that the apartment was too small for us, so installed a tenant there, and rented a more suitable flat for ourselves.

Rahul's plan was to study medicine at university. His qualifications would have allowed him to enter medicine at any UK university, but were not recognised in New South Wales, where the procedure for every candidate for a medical course was to sit the Higher School Certificate examination common to all students, and to obtain a result in the top two per cent of all students in the State. This he did by completing a year's crash course at the local Technical and Further Education (TAFE) College. I had never seen Rahul so dedicated to study as he was doing this course. He would even flatly refuse his mates' invitations to their various parties. His efforts did pay off and on the strength of his results he was successful in gaining admission to the Degree course in medicine at Sydney University – he did not want to study at Dad's university!

****

Although my new colleagues at UNSW frequently invited me to social events, for many weeks after arriving in Australia I felt homesick, and this temporarily affected my mental stability at the time. I needed something to remind me of home—England, that is—to give me some sense of balance. Since my lecture notes had to be prepared from scratch, this occupied my working days but I decided to get involved with the University of New South Wales Rowing Club. This move meant that my weekends were fruitfully engaged. I met a student who was an able rower, Steve Norman, who introduced me to the other rowing club officials. One of them was Murray Clarke, who was an ex-student and dedicated to the wellbeing of the university's rowing club. I was not only rowing myself, but was also engaged in coaching novices, and eventually I formed a crew of Fours. This crew had two members from Mining Engineering and two from other faculties and, in time, achieved creditable success.

At the time, the active members of the club used to meet once a month at each member's home in turn. As a Vice-president of the club, I used to attend these meetings quite regularly. On one occasion the meeting was held at the home of one of the mining students, Richard Van Laeran, and I met his parents, Marie and Van, who became lifelong friends.

The rowing club obtained funding to buy a new scull and my involvement with the club was recognised by christening the boat "Gour Sen" in a small ceremony at the club. (Photo 28)

Photo 28: "Gour Sen" boat naming ceremony at the
UNSW Rowing Club in 1980's

When, after a while, I began to find it very difficult to carry the boat to and from the shed to the pontoon via a very steep ramp, the Mining Engineering workshop very kindly made a trolley to help me carry it. (Photo 29)

A young member of our university's administrative staff,

Photo 29: Scull carrier

Steve Carney, told me that he was interested in taking up rowing. He had been seriously involved in rugby, but after an accident while playing, he took his doctor's advice to give up rugby. Now he was very keen on rowing, but he had had no previous experience. I decided to coach him and took him out in a Pair. Steve took the stroke position and I was at the bow so that I could watch and instruct him. It was a nightmare! He had no feel for rhythm, so we often had to stop and try to find the way to propel both sets of oars through the water at the same time, in rhythm. Eventually he got the hang of it, and progressed very well from then on.

Later, Steve decided to do a postgraduate course at Oxford, where he took up rowing seriously. On one of our overseas trips, I had the opportunity of rowing with him in a Pair in Oxford where he had invited Brigid and me to stay with him. (Photo 30)

Photo 30: Gour (stroke) and Steve Carney (bow) rowing in River Isis, Oxford

We not only had the experience of getting to know this famous university town but were also invited to attend a formal dinner at Steve's college. (Later Steve married the Danish girl he had met who was completing her doctorate at Oxford, and they then went to Cambridge where Steve completed his PhD degree.) (Photo 31)

Photo 31: (L t R) Anna (Steve's wife), Steve, Gour and Brigid just before the Oxford College Formal Dinner

After many years of rowing, on one occasion when I was sculling I realised that the boat was taking in water very quickly and that my seat was already submerged. I had no option but abandon ship, as it were, and swim to shore. It wasn't possible to prevent the scull or its oars from sinking, as it had happened so fast. The following day a diver was employed to do an underwater search in the Parramatta River over some two hundred metres radius from the approximate site, but there was no trace of it; the current had probably dragged the boat much further away, and it was never found. As to the reason for the sinking, it was possible that someone before me had somehow cracked the scull, and that this had not been recorded in the log-book.

A few years later, the UNSW rowing club purchased three boats—an Eight, a Four and a Double Scull—which were christened at a formal dinner function on the university campus to which a number of dignitaries were invited, including the Vice-Chancellor, John Niland. These boats were named after the Vice-Chancellor, the President of the rowing club and me, respectively [see photo]. (Photo 32)

Photo 32: Boat naming ceremony at UNSW. (L to R) Gour, Linda Jansen (UNSWRC President), John Niland (Vice Chancellor)

The University's Sports Association also recognised my services to the rowing club by awarding me Life Membership of the UNSW Sports Association, which allowed me to make use of the university's sporting facilities free of charge.

****

On one occasion I took a university car to a regatta in Taree where our club, including a crew that I had trained, was participating. A student had towed one of our boats to Taree using his car, which, I discovered after the regatta, had now broken down. We couldn't leave the boat there unsecured, so I agreed to tow it on a trailer back to Kensington, well over three hundred kilometres, behind the university car— which, luckily, had a towbar. As I had never towed anything behind a vehicle before in my life, I was very nervous for the whole journey and was much relieved to arrive in Sydney with everything in one piece.

****

I continued my involvement with the rowing club even after my retirement from the university until an incident in 2002. I was rowing in a scull when I suddenly realised that one of my oars had come off the gate. The inevitable result was that the boat rolled to one side and I was in the water. This situation is not uncommon; the standard move is to turn the boat the right way up, then hold both sides and heave yourself onto the seat. I had done this many times in my life before, but the last time I had done so must have been at least five years earlier.

This time, however, my strength wasn't equal to the task. I had to wait in the water holding the scull so that it did not float away, as I could not swim strongly enough to drag the scull to shore. It was May, approaching winter; the water was fairly cold, and I was shivering.

About twenty minutes or so passed, when a "tinny", a small steel-framed boat, came near me, and someone shouted, "Are you all right, mate?" I shouted back, "No, I'm not. Come and help me!" One of my rescuers held the scull while I tried to somersault onto their boat, but by now I was further weakened. The tinny was small and could easily capsize, so they couldn't reach over the side to lift me in; after several attempts I somehow managed to roll over into the boat, and both the scull and I were safe.

As I rolled into the boat, however, I hit my left knee hard on the steel gunwale. This knee had been injured several times before, and this was the last straw. It was so painful afterwards that an arthroscopy had to be performed and about a third of torn cartilage removed from around my knee. After that, I had to take extra care if I were to avoid knee replacement surgery—so it meant that it would be most unwise to do any more rowing on my own!

10

Teaching and Research

The best part of having an academic position was having opportunities to do research work and explore the various blind spots that had bugged me during my time in industry, where the goal was always to finish the job safely and quickly, and to make money. I also had several postgraduate students to supervise, and I allocated a number of mutually interesting projects to them.

Because of my long experience in the industry before taking up my academic position, I was able to squeeze some research funding from industry sources. This also allowed the students to conduct experimental programs at various mine sites to obtain data which they had to analyse to obtain meaningful results for their theses. This was a "win-win" arrangement that both benefited both the students and the mines concerned. To facilitate the on-site research, I periodically had to visit various mines and quarries, through which I became known in industry circles, and which in turn led me to do consulting work that often complemented my personal research interests.

In teaching the blasting course to the undergraduate students, I introduced for the first time a practical component whereby the students went to a mine to see a blasting operation. They were given hands-on experience of drilling holes in the rock, charging them with explosives and finally firing the charged holes. The quarry manager relied on my expertise, and allocated a special area to perform this operation without any supervision from mine management. This practical component was repeated every year during my time at UNSW.

****

At the end of my fixed-term period of three years, in order to re-fill the position I was holding, the university was obliged to advertise worldwide. By this time Brigid and I were quite happy to commit to remaining in Australia indefinitely, so I applied for the position. This involved an interview panel, but I was quite established at the mining school by this time and obtained the tenured position without difficulty.

At this point we realised that it was time to apply for Australian citizenship, otherwise every time we travelled overseas we needed a permit to return to Australia. This was a complicated procedure but in due course we were granted citizenship, making overseas travel less troublesome.

****

The SSP program (or sabbatical leave, as it is called at most other universities) is a program whereby academic staff members are eligible to apply for leave of absence of a specific duration provided they can satisfy the university that the time spent will be beneficial to his or her teaching or research activities, as well as to the University. At that time, the leave period was calculated as one month's SSP leave per semester of teaching. During my fifteen years of teaching at the university I was able to take four SSPs, during which I visited Austria, Brazil, Canada, Czech Republic, Finland, France, Germany, India, Japan, Nepal, Poland, Russia, South Africa, Sweden, UK, USA and Zimbabwe, and at other times I was also invited to give lectures and/or to provide consulting services in Brazil, China, Indonesia, Peru and South Africa. I will mention here some of the more interesting incidents that I witnessed during these overseas visits.

One day in early 1984 I learnt that the Director-General of the South African Chamber of Mines (SACM), Professor Miklos Salamon, would be visiting the school. Miklos had been a contemporary of mine when we were postgraduate students in the early 1960s in Newcastle upon Tyne.

It was good to see him after more than two decades. When he heard that my special interest was in blasting, he mentioned that about twenty per cent of South Africa's gold production was lost due to incorrect blasting practice; there was already a team of researchers busy looking into the causes. He invited me to join them in South Africa. My first SSP was coming up in a few months and although I had already planned to spend some time in the UK, and had just begun discussions with another overseas university, my itinerary was not yet complete; so my first instinct was to seize this chance. Then I thought about my questionable status within the official apartheid policy of South Africa due to my Indian origins. I said, "Miklos, have you thought about the colour of my skin?"

"Oh shit!" he said. Then, after a pause, "No problem, you will be staying at my house."

Well, that was encouraging, but when I told Brigid she was less than enthusiastic. As she was English, she was aware of the hostile reception a mixed-race couple might receive under the South African apartheid regime, knowing as well that at the time co-habiting between whites and non-whites (which included Indians) was unlawful.

Her response was, "You go ahead, and if things turn out well, I'll join you at the end of your stay for three weeks' holiday."

And so I went to South Africa on my own for about four months while Brigid stayed on in England.

PART 6

Travel to Various Parts of the World

11

First Sabbatical

I took my first SSP in December 1984 to return in the following July.

In due course I arrived at Johannesburg airport where I experienced the power of the South African Chamber of Mines (SACM). I had passed through Immigration and was waiting to be inspected by the Customs officer, when I saw Miklos right there—how on earth had he managed to enter this highly restricted zone? He just grabbed my luggage and we walked out without being stopped or questioned, an example of the long arm of the SACM there at the time.

The Salamons lived in Northcliff, an exclusively white suburb of Johannesburg, and they allocated me the guest annexe of their house, but I joined them for most of my meals. (Photo 33)

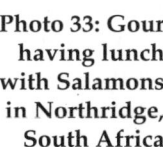

Photo 33: Gour having lunch with Salamons in Northridge, South Africa

I hired a car to get to my office at the SACM and for mine visits, and so on. It took me a week or so to get to know the environment and the staff at my office. After gathering information on the blasting practices in the mines I drew up a plan of action for my research work.

It soon became clear to me that I was going to have a problem if I were to visit mines to see the underground rock conditions for myself. Under apartheid, each mine had separate facilities for changing and washing—blacks only, and whites only. Where was I supposed to change for an underground visit? It would be inappropriate for the mine authorities to allow me to use the "blacks only" facilities; on the other hand, the Boer officials might go on strike if I attempted to use the block for whites. I therefore had to operate by remote control, relying on a white ex-Mauritian mine official, Phil, to collect data for me while he was gathering some other specific data. Unfortunately the results I was getting in this way made little sense, and I became more and more frustrated; my stay there was not producing any useful outcome.

In the end I decided to quit the project and leave South Africa again as I was not occupied in any kind of useful work. When I mentioned this to Miklos he asked me to wait, he would try to sort the problem out. A couple of days later I was told that I would now able to visit the mine and go underground, changing and showering in the mine manager's personal, and quite luxurious, washroom.

These underground visits were certainly very fruitful. By observing the blasting technique in person I realised that I needed to do some experiments on the surface to find the most effective mode of operation. I contacted the local explosives manufacturer (AECI) and explained my plan to them; with their full cooperation, an experimental site was prepared together with a supply of materials for the tests.

Photo 34: Gour in an underground gold mine in S Africa

The experimental results were useful for the small-diameter blastholes used in South African underground gold mines, and were eventually presented at an international conference and later published. (Photo 34)

12

Facing the Apartheid System in South Africa

Before my trip to South Africa I had been in the UK for a short period to observe the procedure for drilling for oil underground. I took several photographs, and in South Africa on one occasion I was showing them to Phil, the white Mauritian mine official. He was clearly puzzled at seeing the drillers unscrewing the drill rods, which was a strenuous procedure. He exclaimed, "Why are white men doing this manual work?" Obviously that kind of manual work was for blacks only in South Africa at that time.

On one occasion I was working in a deep underground gold mine where there was a strict timetable for personnel travelling in the mine shaft so that the shaft could be used efficiently for bringing ore to the surface. The main personnel winding periods coincided with the time of the black miners' shift changes. Often the shaft was quite a long way from the face where they were drilling, blasting and loading the ore. At the end of the shift it was usual for the workers to wait at the plat—the area next the shaft—for the cage to arrive and take them up to the surface. Because of the large number of miners, the cage had to come down a number of times to transport them all and there was always jostling to get into the first cages.

The mine supplied work clothes in different colours. White men wore white overalls, and the colour coding for the blacks depended on their level of responsibility. I was an invited guest, so I wore white overalls.

One occasion, after completing my underground trials, I needed to go to the surface at the same time as the black workers ended their shift. To protect me from being squashed by the rush of blacks when the cage doors opened, a very big Afrikaner escorted me. When we arrived at the plat the black miners were already waiting beyond a rope line. We, being white overall-wearing officials, were allowed to cross the rope and wait near the shaft. There was a bench nearby where my Afrikaner escort asked me to sit while he checked on some machinery near the shaft. A black supervisor crossed the rope and sat beside me. When the Afrikaner saw this he charged across and, shouting abusive language, practically kicked the black man back to the other side of the rope. This inhuman act so disturbed me that I turned my eyes away, feeling helpless in the face of this kind of behaviour.

One day Agi and Miklos took me to see an excellent play at a local theatre; not long afterwards the film of E.M. Forster's classic novel *A Passage to India* was showing at the local cinema in Northcliff, so to return so the favour, I invited them to see the film with me. There was a short embarrassed silence. Then Agi told me that the cinema was for whites only. I was somewhat taken aback, as I had been allowed to see a play only a few days before.

She replied, "Well, tickets to the theatre are more expensive than the movies, and that's enough to keep out the blacks." Dear me!

I wasn't going to let this stop me. The next day I phoned the cinema from my office, and asked the receptionist to connect me to the manager, a Mr Green. She asked for my name and credentials. There was a short delay; since I had mentioned SACM, she must have checked the phone number and after having satisfied herself of the authenticity of my call, she connected me to the manager. After explaining who I was (including my Indian origins), our conversation went along these lines:

Me: I would like to see the film currently showing at your cinema. Moreover I wish to invite the Director-General and his wife, who happen to be whites.

Mr Green: Unfortunately the law of the land does not allow you to be our patron.

Me: There are always laws, but there are also always exceptions. If I am not allowed to visit your cinema, I shall tell the media about this episode when I get back to Australia.

There was a short pause. Then:

Mr Green: I don't have the power to override the law; but I will contact my boss and will come back to you shortly with his decision.

After about fifteen minutes, Mr Green rang me and said, "You are welcome to our cinema. Could you please let me know which day and time you intend to see the film?"

After discussing this with the Salamons, who could not believe my success at first, I rang Mr Green back and told him: "This Saturday at 6 pm."

This was the most popular time and long queues usually formed for this session. Mr Green asked me not to stand in the queue with the others but to report straight to the office, where the staff would be briefed beforehand about our visit and would show us to our seats.

Agi and Miklos were rather apprehensive about this plan. As we approached the cinema we noticed that, sure enough, there was a long queue for tickets.

When we went in, our experience was that everything was well organised, and we thoroughly enjoyed the film. Afterwards Miklos told me that there were a number of hefty bouncers in the cinema that night, something he had never seen there before, and whom he guessed were there to deter any ultra-racist whites in the audience from causing trouble when they saw a non-white in their midst—he commented later that I had managed to open a small crack in the apartheid system.

After several letters to Brigid saying that there would be no problems if she came to South Africa, she asked me as a joke, to go for a walk on the streets with a white woman: if we weren't harassed, she would come to South Africa!

I approached a former colleague from UNSW, Huw Phillips, who had left there to take up a chair at Witwatersrand University in Johannesburg. I asked him if I might borrow his wife, Joan, for an hour or so. When there was a silence and a puzzled look on his face, I realised that perhaps "borrow" wasn't the ideal word, and hastened to explain that my request was not what he possibly thought, but for a special purpose (which also may not have gone down too well); I then explained Brigid's request. Huw was relieved—maybe he was thinking of calling the police—and my request was granted. Thankfully, as I had predicted, there were no incidents during my stroll along the streets with Joan.

Brigid duly arrived in Johannesburg, and by chance the South African Government lifted the ban on mixed-race cohabiting on the very day of her arrival, so we didn't have to sleep in separate rooms!

I was required to audit a coal mine called Kilbarchan Colliery in Natal. Brigid would travel with me by car and stay in a hotel overnight. She would stay in the hotel while I visited the colliery, and then we would drive back to Johannesburg.

While I was talking to the mine manager, a Mr H. Badenhurst, he told me that the mine officials had organised a lunch in their staff club after my underground visit. It was quite evident that they had few opportunities for this sort of occasion, I could see that they would be most disappointed if I didn't accept the invitation. When I mentioned that my wife was waiting for me at the hotel, he immediately said not to worry and that she was also invited, and he would send someone to collect her from the hotel.

So I got changed into work clothes and, duly escorted, I inspected the mine workings and talked for while with the mine officials. After that, the whole entourage went to the club where a large crowd had already gathered, all whites bar one or two, busy getting stuck into the alcohol. Obviously I joined in with great gusto! But where was Brigid? (Photo 35)

Photo 35: Kilbarkhan Colliery Staff Club Bar, South Africa

She later told me that, when she heard from the hotel management that someone was coming to fetch her, she waited in the hotel foyer. She saw a sari-clad woman come in, obviously looking for someone. After some time Brigid realised that she herself was probably the "someone" in question, so she went to her and introduced herself as Mrs Sen. Immediately the penny dropped; the Indian woman was not expecting Mrs Sen to be white.

By the time Brigid arrived, I had far more to drink than was safe to drive on to Johannesburg that afternoon. The party was fabulous, and the discussions were so frank that there were no inhibitions on either side. We talked about apartheid quite openly and, of course, some of the men there were completely dogmatic when justifying their racist policy.

There was one Indian man at the club, the manager of the black workers' hostel, but he was only there because his wife had been Brigid's escort from the hotel. I later learnt that this was the one and only time he had been allowed into that whites-only club!

When everybody realised that we were not travelling back to Johannesburg that day, the party carried on at full steam until the club's closing time. Then the Indian manager invited all of us to his hotel to continue the party. Everybody was in such a happy mood that most of them went along. The attraction was largely the discussions, which focused mainly on society and politics, two topics that are usually avoided in polite company. Afterwards my wife told me that she was talking to a white mine official who asked her why she had not married a white man. The less said about such a level of racism, the better, I think.

****

Agi and Miklos invited us to visit Kruger National Park with them one weekend. This was a wonderful experience: living in a thatched hut, a rondavel, in the grounds of the park, but well protected from the animals. We saw impala, monkeys and other animals, but I thought the most entrancing were the elephants, with their majestic movements. (Photo 36)

One lunch time we were having our sandwiches and drinks in a protected enclosure when I recognised a young man approaching us.

Photo 36 Salamons and Sens in Krugar National Park, S Africa

By an extraordinary
coincidence it was
Kent Smith, whom I
had last seen as a
schoolboy when he
was a student of mine
at Wynstones School
back in England
nearly ten years
before. Kent was now
married and living in

Photo 37: Sens with Kent Smith's family,
Cape Town, S. Africa

Cape Town. This visit to Kruger National Park was his first,
as was ours. How strange it seemed to meet in a completely
strange environment when our homes were on different
continents! He invited us to stay with them if we were ever
in Cape Town, which we did soon afterwards. (Photo 37)

Kent had designed and built a marvellous hexagonal
house, where we stayed for a couple of nights. He told us his
work, which from his description sounded very challenging,
was connected with local non-white youths, involving them
in various constructive projects.

****

Since Johannesburg is on a plateau, it has no rivers; but
rowing enthusiasts were not about to be cheated of their sport
by a mere geographical accident. Wemmer Pan, a large
settling pond about a kilometre across—once used by gold
mines, now closed—amazingly had no fewer than four
rowing clubs on its shores! (Photo 38)

Photo 38: Wemmer Pan home of 4 rowing clubs, about 1000m diam.
Jo'burg's old settling tank, S. Africa

13

Rowing in a Pond

I rowed regularly when I was in Johannesburg, and even joined in a team regatta one Sunday. The regatta course was from one corner of the pond to the other, and so was barely a kilometre long: by the time we had taken a racing start followed by about ten strokes, we had reached the finishing line. As per the norm, after the races we all settled down to drinks and barbecue with a bit of chitchat—all very enjoyable.

One point I should mention here is that I didn't notice any other non-white rowers during my stay in Johannesburg. Was this because of the colour bar, or due to lack of interest, or some other reason? I didn't enquire.

My two-week holiday was approaching, so we started making arrangements to see some other parts of South Africa. One weekend we thought we would drive to see the famous bird sanctuary in St Lucia. At that time the St Lucia Hotel was the only one with accommodation to European standards, so I sent them an urgent letter asking them to reserve a room for us, making sure to mention our details (that I am of Indian origin and that my wife is English) in order to avoid any problems on arrival. We didn't receive a reply for some days, and the Salamons said that might mean that the hotel for whites only, and didn't bother to respond for that reason.

We were about to make other arrangements, when a letter arrived from the St Lucia Hotel confirming our booking. In due course we drove to St Lucia, but by the time we got there it was well into the evening and difficult to see the road signs, so we stopped at a shop and asked for directions to the hotel. The woman, who was white, was a bit surprised by the question, but she indicated the direction of the hotel. In passing she mentioned that the hotel was for "whites only"; I ignored her remark and drove to the hotel.

Upon arrival I asked Brigid to stay in the car while I went to the reception desk to check in. I saw straight away that the foyer was filled with whites; suddenly all conversation stopped and everyone's eyes were upon me.

I told the receptionist that I had a room booked, and showed her the letter. The poor woman was at a bit of a loss, but quickly composed herself and said, "Please wait, I'll fetch the owner." When the owner came he could not have been more polite, and welcomed us to his establishment. We didn't want to know if there were a backlash to our stay at the hotel or not, but we did not experience any problem whatsoever in our time there, and thoroughly enjoyed ourselves at the hotel and at the wildlife sanctuary.

After our visit to St Lucia we headed towards the mountains where we could get a bird's-eye view of the region from high above the plain. We then saw a signpost to a hotel and rather liked the idea of popping in for a cup of tea. I waited in the parked car while Brigid went inside to see if it would be possible. She came back a few minutes later and gave me the bad news. She said that the receptionist was very welcoming at first but as that as soon as my background was mentioned, the receptionist had said, "Sorry, this is not an "international" hotel (where all races are allowed). If you drive down the road you'll come across one." So much for our first blatant racist experience together—although I had had some unpleasant moments.

****

During our stay in Johannesburg we made several friends, a well-established local family, Dot and Brian Zylstra. We were introduced to the Zylstra family by the Nells whom had we met earlier in Simla, India. The

Photo 39: (l to r) Dot, a guest, Brigid, Brian and Gour in Zylstra's house, Jo'burg, S Africa

Nells and Zylstras had been friends for a long time; Brian was a successful businessman with international connections. Some years later they emigrated to Australia with their three sons, while retaining a base in South Africa to look after their business there. When Brian heard that we were contemplating a visit to Cape Town, he suggested that we get together with one of his contacts. (Photo 39)

We met Brian's acquaintance, who took us to a winery for a wine tasting. After that he entertained us in a rather posh restaurant for dinner where the concierge noticed that I wasn't wearing a tie, which contravened their dress code. So, what to do? Our host asked for the manager, who very kindly lent me a tie and the problem was solved; I suspect it wasn't the first time this had happened—frankly, I would never have anticipated needing a tie, especially here in Africa!

The next day we took a cable car to the top of Table Mountain and took in the breathtaking view of Cape Town and surrounds. On the mountain itself we saw some dassie, or rock hyrax, running about—small rabbit-like animals that are unique to that area.

At the end of our Cape Town jaunt we took the world-famous Blue Train for the two-day return trip to Johannesburg. This trip had to be booked well in advance. It was luxurious rail travel at its best. We were so pampered that I felt that I was in a sumptuously furnished five-star hotel. What a different journey from our awful experience when my whole family travelled from Bombay to Calcutta

Photo 40: Dining in "Blue Train" on Christmas Day 1984; Brigid and Gour is in far left and right; S Africa

back in 1962! As far as I could see, I was the only non-white passenger; in fact, even the railway staff were white, including the catering and cleaning staff, which was unusual. Every meal was served in the well-appointed restaurant car, and the food was of the highest standard. (Photo 40)

It was said that the carriage windows were impregnated with gold dust so that the sun's rays were reflected outwards to keep the carriage cool. This journey will stay in our hearts for the rest of our lives.

****

My Promotion

The year following my return from South Africa, I felt that I had made sufficient contribution to the Mining School to apply for promotion. I had to do this very carefully, submitting a document with my application that showed my achievements over the past four years for scrutiny by the interview panel members, followed by an interview. I was successful, and promoted to Senior Lecturer in 1985.

This promotion involved my accepting more responsibilities in the running of the mining school. I became the Director of Undergraduate Studies, which involved scrutinising the curriculum for undergraduate students as well as organising their field trips and vacation practical experience, among other duties.

14

Sad News from Tokyo

Sadly our eldest son, Prasanta, who had been living and working in Tokyo, was killed in a motorcycle accident in January, 1986. This news was so shattering to Brigid and me that at first we were stunned by it, but arranged to fly to Tokyo where we were met by Fumiko, Prasanta's ex-girlfriend, as well as by a very good friend of his, the author Mike (Masahiro) Miyazaki. We were taken straight from the airport to the hospital where our son's body lay in a dignified manner on a bench surrounded by flowers and burning candles. We were crying inwardly at seeing him thus. I am sure there were some tears, but we had to restrain our emotion when in Japanese society.

Afterwards we went to Prasanta's rented apartment in the Roppongi district of Tokyo, where we stayed for two weeks until the end of the tenancy. One after another, Prasanta's friends and colleagues came to support us in our vulnerable state. From his friends we also learnt all about his activities in Tokyo. Besides giving lessons in English, he provided translation services, played the saxophone, along with many other things. Prasanta was also working part-time for Toshiba Medical Engineering Company. It so happened that on the evening of the motorbike accident, Toshiba's Board members had decided to offer Prasanta full-time employment with the company, something almost unheard of for a non-national and therefore a true honour; again, sadly, this exciting news never reached Prasanta. The president of his company presented us with a photograph of him taken a few weeks before his death, with a very kind message. (Photo 41)

In memory of Mr. P. Sen,   February 10, 1986
You have done an excellent job for us, and have been loved by
all of us, like our brother. May God bless you forever.

President, Toshiba Medical Engineering Co., Ltd.

Photo 41: This photo of our Son with a short message presented to us by President, Toshiba Engineering

Prasanta's friends did all the running around with regard to completing the police report, making funeral arrangements and inviting people to the funeral. We are extremely indebted to these friends who rallied round during our time of distress in a totally foreign environment. The funeral took place in Zo-jo-ji Temple, in the heart of Tokyo, in accordance with Buddhist traditions, and an urn with our son's ashes was deposited in a vault within the temple grounds. (Photo 42)

Photo 42: Our son's last rite at Jojo Ji Temple, Tokyo, Japan

Our son's death was the catalyst for a new door to be opened for us in Japan, where we made a strong connection with very many friends. Eventually I became professionally involved in Japan on many occasions thereafter.

15

First Experiences of China

In the early 1980s the UNSW School of Mining Engineering was attracting a large number of postgraduate students from China. In 1987 one of my research papers was accepted at the International Mining Safety Conference in Beijing, and Brigid decided to accompany me there. When they heard that I was going to the conference, two Chinese students asked me if I would deliver seminars at their institutions. Apart from Beijing University, where the conference was being held, I agreed to two of their proposals for seminars to take place at universities south of Beijing, provided that their universities would pay our internal fares and hotel expenses, but no lecture fees. We planned to then return to Sydney, stopping in Hong Kong on the way.

After our arrival at our hotel in Beijing, we received a delegation from one of the southern universities. They announced that the government would not allow them to buy flight tickets for us because we were there to attend a conference and not to present seminars. This, of course, put us in a very difficult position. The arrangement made previously in Sydney was that all our flights as far as Hong Kong would be provided for us. I couldn't afford these flights myself, and by this time it would have been almost impossible to get on a flight at such short notice since most of them had already been booked for the government officials' travel; meanwhile the delegation was pressuring me to agree to visit their universities. This so frustrated me that I refused to give seminars anywhere, not even in Beijing.

The spokesperson from Beijing University came to our hotel and begged me to just visit their mining department. In the end I agreed, and went to the university to meet a roomful of faculty staff. After the formal introductions were over, someone told me that my proposed seminar had been announced in their magazine, and delegates from all over China had arrived and were waiting for me in the lecture hall—would I be kind enough just to say hello to them? Otherwise they would be very disappointed. What could I say? I agreed to do this.

When I arrived in the hall it was completely full; by my estimate there were close to a hundred people there. The interpreter asked me to explain my position and my research interests to the crowd. After I had done this, the interpreter said a few words. There was then a dialogue in Mandarin between some of the participants and the interpreter, who then asked me: "Would you mind answering some of the attendees' questions?" I had been trapped, of course, but I had to honour my profession, so I agreed. The questions and answers went on for an hour, and the content more or less covered the intended seminar theme. That was the Chinese way of doing business.

Now we had to find a way out of China! I scanned the details of the various post-conference technical visits and found that one of them would terminate in Guangzhou (previously Canton) after visiting a number of sites. This might do it, I thought, but how was I to pay for it? The conference officials would not accept credit cards, foreign currency or personal cheques; it all had to be in Yuan. I knew no-one there from whom I could have borrowed the whole amount to pay for both our fares. Eventually, after borrowing money in various currencies from friends and acquaintances, I converted it to Yuan and handed it over to the officials—

who, incidentally, were exceedingly unfriendly throughout our negotiations. I ground my teeth throughout the whole procedure and vowed never to set foot in China again.

Some of the post-conference technical visits were informative; it was quite clear, for example, that safety standards in the mining industry were very low. Miners working underground did not wear safety boots, nor did they have hard hats to protect their heads from rock falling from the mine roof. It was not surprising that so many safety-related mine accidents occurred in China.

After one underground visit we were treated to lunch. Four guest couples were seated at each table, together with two mine officials. During the mealtime we were offered special bottles of Maotai, China's national alcoholic drink. (Photo 43)

Photo 43: Lunch after visiting an underground coal mine, Technical Tour, China

It seems that Maotai had been the favourite tipple of Premier Zhou Enlai. He wrote a poem about it, which was printed on the label; I asked our host if we might take an empty bottle as a souvenir. He immediately ordered a full bottle for each of the couples at his table. Hosts at the other tables saw this, and—perhaps obliged by some sort of tradition—did the same for their guests. Then our host, the most senior member of the mine staff, wanted to show off his power and ordered another round of bottles for our table; this was not copied at the other tables, however. By the end of lunch, Brigid and I each came away with a bottle of the Chinese national plonk.

****

During one of the post-conference excursions, the bus in which a group of a dozen or so delegates were travelling broke down. Our driver told us that we were going through a prohibited zone—meaning an area that the People's Republic did not want foreigners to see—and that we must stay in the bus; another was on its way and would be there soon. The broken-down bus attracted all the locals, who started crowding around us. It was a hot day; after more than half an hour sitting in the bus, waiting, we persuaded the driver that we would be ill if we didn't get some fresh air. Naturally, as soon as we got out we walked around and saw for ourselves the poor living conditions there. (Photo 44)

Photo 44: Our technical tour bus broken down in restricted zone attracted large crowd, China

16
Highlights of our Travels

In Groote Eylandt

Groote Eylandt, owned by the Anindilyakwa Aboriginal people, lies fifty kilometres from Northern Territory mainland, and is the largest island in the Gulf of Carpentaria. The Anindilyakwa currently lease the rich manganese orebody on the island to the mining company, currently GEMCO, a BHP Billiton subsidiary, who pay royalties to the Anindilyakwa people.

There are regular commercial flights between Darwin and Groote. At the time of my visit, however, the island was not open to tourists and access was by permit only (this is no longer the case). I was visiting at the invitation of the mining company so Brigid and I had no difficulty obtaining permits. We stayed at the mining company's very well appointed guest house. (Photo 45)

Photo 45: Mine guest house where we stayed, Groot Island, Northern Territory, Australia

In Coober Pedy

Ann, the mother of one of my students, owned an opal mine in Coober Pedy in South Australia—the largest opal mining area in the world—famous for its "dugout" underground homes. She invited Brigid and me to visit her mine, on the condition that we accompany her to the mine early one morning and stay there for the whole shift of about seven hours. I was happy to do that as long as I could be involved in some sort of simple task. It would be a nice change; I hadn't worked an entire shift in a mine for many years!

Getting out of the air-conditioned car in Coober Pedy we could feel the intense heat, so went straight to our hotel, the Desert Cave, which was also dug into the rock with the rooms under the ground, which kept the temperature there pleasant and more or less steady throughout the year. (Photo 46)

Photo 46: Underground Hotel in Coobar pedy where we stayed, Australia

When Brigid saw the mine shaft—a typical Calweld auger-drilled shaft about five metres deep and about a metre in diameter, with an iron ladder to climb down—she refused to climb down it but said she would stay in the car and read the book she had fortunately brought with her, or just walk around from time to time. On the surface, a tent covered a conveyer belt carrying the waste rock, or mullock, from the mine. An old lady was sitting in the tent, picking out any stray opal off the belt. Brigid visited her a couple of times for a chat.

I was fascinated to see the mine workings, which were quite mechanised. While a man operated a mining machine similar to a rotary-head tunnelling machine, Ann "spotted" for opals, and cleared the broken rocks using a blower that operated rather like a household vacuum cleaner, used in reverse. (Photo 47)

Photo 47: A mechanical Miner cutting out, Coobar Pedy, Australia

Broken rocks were dispatched to the surface by a continuous telescopic pipe. I occupied myself adding extra length of pipes when required. In slack time I looked around the workings, which comprised a series of horizontal drives about twenty metres long, radiating from the shaft.

At the end of the shift Ann had three small bags of opals, and she was so delighted that she donated three pieces to me. She thought I must have brought them luck—they hadn't found any the previous week—so could I please visit again the next day? (Photo 48)

Photo 48: Underground chambers of a opal mine, Coobar Pedy, Australia

In Uluru

The first leg of the journey was by train from Sydney to Alice Springs. After a few days in and around The Alice, we went by coach to Uluru (or Ayers Rock as it was still called then) in the Uluru–Kata Tjuta National Park—an enormous monolith, or single piece, of some of the oldest rock on Earth, nine and a half kilometres around and rising nearly three hundred and fifty metres above the surrounding plain. We spent a couple of days there to enjoy the "emptiness" of that region.

One day we went to climb The Rock, which was a popular tourist activity[3]. It had been drizzling and parts of the rock were slippery. When we had climbed a little way there was a resting place with benches, and Brigid was reluctant to climb any further but since I was determined to go further she agreed to join me. Due to the slippery rock, certain parts of the walk were somewhat risky. When we arrived at the summit it was a great feeling of achievement. The scenery was fabulous. (Photo 49)

Photo 49: (top) Brigid & Gour at the top of the Ayers Rock

3. Uluru was handed back to its Traditional Owners in 1985. Various proposals have since been put forward to ban climbing on this ancient Aboriginal sacred site for cultural and ecological reasons, but the Australian government has always blocked such attempts. Meanwhile, the Traditional Owners ask visitors to be sensitive to their wishes and *not* to climb. Some people do respect this.

At the summit we saw a well-documented yet still strange phenomenon: a small depression filled with recent rainwater, forming a pond containing small fish. Most of the time there is no water, so ... how do they survive for months without water?

Climbing down was trickier than climbing up. When we arrived safely back at ground level we saw a notice board. Brigid said that if she had seen it before, she would not have even attempted the climb—she has fear of heights! (Photo 50)

Photo 50: Warning sign advises who should not climb Ayres Rock

In Indonesia

I had an assignment to provide a blasting course for the engineers of a newly commissioned open-cut coal mine in Kalimantan, Indonesia.

As per my previous trips I wanted to have my wife Brigid along to accompany me on this trip but I was not able to obtain the necessary visa for her to enter Kalimantan. It appeared that, due to the political sensitivity of the area, perhaps because the location of the mine was very close to the borders with countries with which Indonesia had uneasy relations at the time, the only persons allowed to visit Kalimantan were those whose work schedule required their presence. This refusal was a great disappointment to Brigid.

I flew to Jakarta from Sydney, and look another flight from there to Kalimantan where I was met with a car and driver. After driving a considerable time, I arrived at the mine-site and was escorted to the expatriates' staff quarters. These turned out to be containers, divided into three sections, each section becoming a room which was very frugally furnished with a narrow single bed, a small table and a chair. (Photo 51) This was all the space any staff member could have for sleeping, reading or relaxing, no matter how senior a position that person might hold. Fortunately the room had an air-conditioner which helped me to get a good nights' sleep, but I did think how lucky it was that Brigid had not come—she would probably have been a most unhappy visitor if she had.

Photo 51: Interior part of the staff quarter, Malimantan, Indonesia

One day during my stay the water purification plant failed to work and the resulting lack of running water caused us all great distress. For a while all we had was bottled water to serve all our needs, including washing.

There were 14 young engineers attending my course. (Photo 52) It was very satisfying to note that these engineers were very keen to learn the techniques of safe and efficient blasting. Apart from instructing them in the classroom in the salient features of blasting, I joined them on site visits to supervise and monitor their skills in the field.

Photo 52: Course participants, Kalimantan Coal Mine, Indonesia

Although I lived for the best part of a week in Orang Utan country, I did not get an opportunity to see these very special animals in the wild, which was not surprising, but nevertheless a great disappointment.

At the close of the course I was escorted by an engineer back to Jakarta. When we arrived in Kalimantan Airport we discovered that our plane's departure was delayed by two hours, so my escort proposed that we have a look around in the surrounding area. We went to the local market, where I saw that a stall holder was selling a 'Black Orchid' (Paphiopedilum species). This type of orchid is very rare and Kalimantan is one of their few homes. I was told that the Indonesian government did not allow Black Orchids to be sold in the market place, let alone be taken out of the country.

Orchids are a passion of mine and I was obsessed with the idea of possessing such a rare and delicate specimen. My Indonesian escort told me that if I purchased the plant he would take the responsibility of carrying it as far as Jakarta. It seemed an opportunity too good to be missed, so I did buy it, and decided to take it with me to Australia Accordingly, the orchid plant was handed over to me after our arrival in Jakarta. So far, so good. Then I decided to try smuggling it into Sydney by burying it in my hand luggage.

After our plane touched down in Sydney, as I was walking along the corridor to the immigration section, I caught sight of the huge signs displayed there, warning not only of the huge fines but imprisonment as well that would be imposed on anyone who flouted the quarantine rules. At first I thought I could ignore the sign and take the risk, but after seeing it again and then again, I lost my nerve, imagining not only the huge amount I would be fined if the plant were discovered but seeing in my mind the Headlines: University Mining Department Head Caught Smuggling! I decided to declare the orchid to the quarantine officers, who promptly confiscated it and threw it into the bin. I was partly relieved but naturally disappointed too, and related this incident to my family back at home. A couple of years later our son Rahul gave me a 'Black Orchid' plant as an extravagant and beautiful birthday present.

In Romania

In 1999 I travelled to Romania in order to present a paper at an International Mine Safety Conference. The conference venue was a hotel situated in a beautiful hilly holiday resort called Sinaia in the Carpathian Mountains. As usual, my wife accompanied me. She had read up on the monasteries in Romania with their extraordinary frescoes, and also the possibility of home stays within the country and was very keen to do both. Accordingly, I enquired of the conference organiser whether this could be arranged for us and he eventually succeeded in his request to a local delegate to look after us for about a week during our stay in Romania.

So after the conference we met up with the delegate, Paul Colceru, as arranged, who drove us in his small car to his local town. The journey was a bit fraught; it was very hot and we did not know Paul, nor did he know us, which made conversation difficult. However, we gradually relaxed and began to exchange ideas and information about our families. In fact, we managed to understand each other so well that we became good friends.

We spent the night in a hotel in the town where Paul lived. The following morning Paul met us with his car at the hotel to take us on an exploration of the area, to mineral rich springs gushing out from the hillside where he filled bottles to take back with him. At the end of the day, after Paul had had time to get to know us better, he tentatively suggested that we might like to visit his parents and stay with them overnight; his parents were happy with the idea, he told us. He warned us, however, that his parents did not speak English and that they lived well away from shops etc in their small farm in a village. Did we think we would be able to cope with the situation? But we were delighted with the idea: when would such an opportunity come our way again? We accepted the invitation with alacrity but stipulated that we must give his parents the cost of our hotel room in return. Paul's parents, Petru and Maria, had a wonderful rambling house, not large but comfortable and homely, with a big vegetable garden and sheds for their cows, pigs and goats, with hens roaming freely in the yard and garden with their chicks, a dog that spent much of its time on the low shed roof, tied up and on guard duty, and a cat that had learnt, reluctantly, that the chicks were not to be caught for its dinner. Petru and Maria worked hard every day on their farm, in the steep hayfield up behind the house, tending the animals, or working in the garden at the front, making their own wine and distilling some of it too, a small glass every morning to start the day! Practically the only food items they had to buy were sugar, oil, coffee and tea; in all other respects they were self-sufficient.

So we went to stay with them, wondering slightly how it would turn out. We did not have the luxury of hotel set-up, true, but the comforts of home-produced food, such as honey, butter, garden produce, all cooked for us by Maria in her simple farmhouse kitchen, more than compensated. My wife's knowledge of basic French and school Latin made some communication at least possible, since Romanian is a Latin based language, and we got on famously with the senior Colcerus. In fact we got on so well that the one night's stay stretched to three and even then it was quite a wrench to leave them.

On the third day, after returning to our hotel, Paul met us there to join us for coffee, and we reminded him of the agreement: what we had saved on hotel expenses was to go to his parents. Paul was horrified at the idea and was unwilling even to consider it. For him a guest was an honoured person, to be made welcome and given freely of hospitality. The arguments went back and forth, or rather round in circles as Paul pointed out, but in the end a solution was found: he would accept money as a gift to his elderly grandmother who lived in the city on a pension and had none of the benefits of his parents' country living. Thus honour was satisfied on both sides.

During our stay, Paul drove us to various places in the region so that we could visit a number of monasteries, each one with its particular attraction, the mural paintings, architecture, or frescoes. A most beautiful monastery was

Voronetz, which has frescoes all round its outer walls (Photo 53) and is on the list of UNESCO heritage sites.

Photo 53: Voronet Monastery, Rumania

Towards the end of our time in Romania, we met up with the conference organiser once more and drove with him to a monastery in Suceava County, it may have been Putna but I have no record of the name. This monastery covers a large area and, owing to its relatively secluded location far from Bucharest was not vandalised during the Ceausescu regime, unlike most other monasteries. The monastery has a large library with a huge stock of both modern and ancient books. We were privileged to be able to spend two nights in their guest rooms and to eat our meals in the refectory with the monks, who sat at a long table below a modern painting of The Last Supper in vivid colour, with bright gold halos crowning the saints.

This monastery had also avoided the attentions of Ceaucescu's destructive regime by pretending to be an educational institution and keeping religious things well hidden. There was a large hall, part of the monastery but set apart somewhat, that was in the process of being 'redecorated'. This meant that an artist was in residence and spent his days up on scaffolding while he worked on a huge fresco on the end wall. We stood and observed his work for a while and my wife noticed that his tools and painting kit were parked carelessly on a hand-woven large blue and white cloth. It caused her some distress to see such a beautiful piece of handwork being so abused, and she asked me to enquire whether we could rescue it by buying it from the monks. This caused some consternation, when finally put to the Abbot by our conference organiser host, who was with us on this trip. It was explained that as the bedspread, as it turned out to be, had been presented to the monastery as a gift from a devout worshipper, it could not simply be given away. After considerable and delicate negotiation, the Abbott conceded that it was not being used to its full value and that a donation to the monastery would enable us to take it and care for it. There followed a 'think of a number!' exercise which determined the size of our gift, and the bedspread was ours. In our turn, we requested that the donation be spent on the Boys Home that the monastery was running as part of the operation to rescue boys who had suffered so much dreadful neglect and abuse under the Ceaucescu regime.

On our second trip to Romania, some years later, we went as tourists. First we visited our friend Paul and his family, and then set off on our own to explore more of the country. This time we visited most of the sights, such as Bran Castle of Dracula fame, and the medieval town of Sighisoara, and other places of interest. In Sighisoara we stayed in a hotel which displayed the signed photo of Charles, Prince of Wales, in the foyer: the Prince had visited that hotel during his tour of Romania. Clearly, we had chosen our hotel well! This is a fascinating city, on the UNESCO Heritage list, with an excellent museum displaying weapons and some fine ceramics, which pleased my wife, and has an amazing clock in the Clock Tower. We watched its fascinating mechanism and the little figurines that move at certain times of the hour. This very ancient town is set high up above the river with a steep covered walkway cut through the thick walls of its original fortifications to the main market and shops below, and from the tower on the top we could see for miles round.

Travelling with an African Prince

On one occasion we travelled through various European countries by train on a EURAIL pass, bought before we left Sydney. One day, in the train crossing Switzerland, sitting alone in a first-class compartment and next to the door was a well-dressed African. When we entered we were a somewhat surprised that a man from Africa could afford to travel first class; he must be really wealthy, we thought. I was sitting opposite Brigid, next to the window. To be polite I moved over and sat beside her and indicated to the man (whose name I now forget, but let's call him Mr X) that he could sit next to the window, if he wished. Mr X politely declined and continued to sit next to the door. I found it awkward not to make any conversation with Mr X, so I introduced Brigid and myself to him. After that he relaxed a little; eventually he told us that his father had been a king in Mozambique when it had been a Portuguese colony. When the communists seized power, his father and other family members were murdered. As he was abroad studying at the time, Mr X was safe from the persecution; however, the present regime in Mozambique was still hunting him, and he felt he needed to sit near the door in case he needed to defend himself, or to give himself a chance to escape. He had been given a protection visa by the Austrian Government, and had spent time to Japan to learn martial arts for self-defence.

Was this a true story? We like to think so. He spoke perfect English and his general behaviour suggested to us that his claim to a royal heritage was genuine.

17

Second Sabbatical

During my second SSP leave (July 1988 to February 1989) we spent roughly equal periods of time at Kyoto University (Japan), Bengal Engineering College (India), Leeds University (UK) and Colorado School of Mines (USA), with some brief side-visits to other overseas institutions and mines.

In Japan

During this time the stated aim of my SSP was to familiarise myself with the areas of research work being carried out in various institutions and establishments. I was also invited to present some seminars on the research I had been involved with in Australia. One of the especially important visits was to Hishikari Gold Mine in Kagoshima Prefecture. The mine had the richest gold vein I have ever seen, with the gold content averaging eighty grams per tonne. The gold ore was transported in an armoured truck with an armed security guard to be processed some distance from the mine.

In the mine, the ore occurred in rock that was saturated with hot water (average temperature 65°C). A well-controlled systematic dewatering scheme lowered the water level; the hot water was reticulated through a pipe network and sold to neighbouring households as a source of revenue for the mine.

While I was occupied in the university, Brigid liked to explore Kyoto by public transport, and she became quite adept at interpreting the bus routes from the Japanese signs at the front of the bus. One evening there was a special market being held in the temple area. After dinner we went to the market and took in the large selection of exciting local produce; after a couple of hours, we thought that it must be getting late and that we should head back to our apartment. We boarded a bus that seemed to be going our way, but soon realised that we couldn't recognise anything we passed. Brigid spoke a little Japanese, so we tried to explain with the aid of a map with our apartment building marked on it. The driver looked most concerned, and asked us to stay put.

Eventually the bus pulled into the terminus where all the other buses were parked for the night. The driver took us into the office where some people were milling about. We understood that there were no other buses running at that time, nor were there any taxis nearby. After some serious discussions among the staff there, we were told to accompany a man who invited us to get into his car. We gathered that he was going to take us to our apartment, although he lived in the opposite direction. We didn't want to inconvenience him any further, and said we would be happy for him to drop us off where we could find a taxi. He would not hear of it, but took us right to our door—about a thirty-minute drive. We offered him some money, to buy something for his children, but he vehemently refused to accept it; I hoped we hadn't insulted his generosity by so doing.

We thanked him profusely and waved him off. That is Japanese culture! Can one get that kind of help anywhere else?

At lunchtime at Kyoto University I would sit at the same table as Professor Ashida. My usual practice was to buy a packet of sandwiches from a shop on the way from the apartment to the university. Professor Ashida used to open his bento, traditionally a home-packed meal in a box, wrapped in cloth. A typical bento might comprise rice, fish or meat, usually with pickled or cooked vegetables. I noticed how beautifully the items were arranged, and one day I asked him where he bought it. He said, "My wife prepared it. Every morning when she gets up that is her first job."

Professor Ashida travelled from Osaka, about one hour from Kyoto, and he arrived every day in his office by eight in the morning, so I guessed Mrs Ashida had to get up at five o'clock at the latest in order to prepare the bento for her husband. What dedication! I tried to convince Brigid to take a leaf out of Mrs Ashida's book, but without success.

I tried to be involved with one of Kyoto University's rowing clubs. We went to the club house some distance from the centre of the city. It was housed in a large building with a huge hall where at least 50 boats of various types were stacked in an organised way, and a number of dormitory-style bedrooms where the selected rowers lived, attending university lectures once a week during the regatta season and practising their rowing both morning and afternoon on the other days. All food was supplied during this intensive practice time.

The club was situated on the very fast-flowing Seto River, not far from Lake Biwa, Japan's largest freshwater lake. I watched the students' practice session, which was very rigorous, with a high stroking rate requiring a great deal of stamina. I wondered how well they would sustain this over long stretch of, say, two kilometres. I found out that this kind of training procedure was practiced in every sport.

****

Shikoku is the smallest of Japan's four main islands, and it was not so popular with tourists at that time. Initially the island was connected to Honshu by a series of six bridges, collectively called the Seto-Ohashi Bridge, carrying both trains and cars. Since then the Akashi-Kaikyo Ohashi, or Pearl

Bridge, with the world's longest central span, has been built to connect Shikoku with Honshu.

My main idea was to visit the Torigatayama Quarry Complex limestone mine, which had a very economical power supply system for the main conveyor. However we were keen to experience some of the island's special features, including a youth hostel run by Buddhist monks at Awa-Ikeda. Then we decided to stay at a minshuku (family-run guest house) in a small town.

We had already spent more than a month in Japan and our finances were becoming rather tight, but while we were exploring the town we came across a small bank with only two or three employees, but with a sign saying VISA. At that time credit cards were not common or generally accepted, so it was surprising to see the sign in a small place like this. I went in, enquired about withdrawing some money, and to my surprise I was able to withdraw a million yen without any fuss. We were much happier now that we could splash out on good meals and so on.

After coming out of the bank we were doing some exploring, when a motor scooter stopped near us. When the rider took off his helmet I recognised him as the bank teller who had just given me the money. My heart sank, fearing he wanted the money back—but he explained that he should have asked to see my passport for their records, that was all. Relieved, I quickly went to our minshuku to show him the passport, glad that the temporary scare was over.

Waking one morning in the minshuku I felt itching all over my body. Brigid quickly saw that they were flea bites. Why was I the only one affected? Obviously my tatami (Japanese mat) was infested with fleas, but not Brigid's. The irritation was such that we had to go to a chemist to find some medicine. Then, after mimicking the flea action for him and showing him the spots on my skin, the chemist exclaimed: "*Nomi* (flea)!" and produced a tube of ointment that gave almost instant relief.

****

In Nepal

After leaving Japan in September 1988, we took a short break in Nepal en route to Europe. Apart from my wish to carry out a professional audit of a marble quarry near Kathmandu, our mission was to visit our Plan International-sponsored child; we had taken up the sponsorship in memory of our late son. The child's family all lived, slept, cooked and ate in one room; above the room, in an attic room, they raised chickens to supplement their income.

We hired a car and made some short trips in and around Kathmandu. The driver invited us to have a drink at his local "pub", which was in a dimly lit, low-ceilinged hut. The local brew, *bhaati jaanr*, was a warm fermented rice beverage, and was fairly potent; after a few adventurous sips we decided to return to our hotel.

In Austria

I was invited by Professor Günter Fettweis to visit the Mining Department at Montan University, the smallest technical university in Austria, located in Leoben, central Austria. He met us at the railway station, and what a remarkably distinguished and respected figure in the community he was. Even the shop assistant warmly hailed him with the German miners' greeting, "Glück auf!"

Our visit coincided with the university's annual *Ledersprung* ceremony when the first-year mining students and new faculty staff are initiated with mediaeval pomp. When we arrived for the ceremony, the hall was filled to capacity. Alumni were also there, very happy, with much loud talking and laughter—perhaps a goodly amount of alcohol had been consumed beforehand. We were ushered to our seats in the distinguished visitors' row.

There is no boundary between the faculty staff, students and alumni. The solidarity of the mining community is emphasised by the clothes—the traditional black jacket, or

Bergkittel, of miners in Central Europe, particularly Austria and Hungary.

The procedure was this: a beer barrel is placed in front of the gathering, and the Rector and the oldest alumnus held a miner's leather apron toward the barrel. Then one by one, the candidates climbed on the barrel while holding a glass full of beer, and some questions are put to them. They then quaffed down their beer, jumped over the apron, then passed under the upraised swords of a guard of honour. (Photo 54)

A student leader asked if I would go through the ritual, as I was a miner witnessing the ceremony for the first time. What could I do but accept his invitation? All questions were in

Photo 54: Gour performing Ledersprung ritual, Leoben, Austria

German, of which I have no knowledge, but all involved personal details—name, place of birth etc., and were asked in sequence. As long as I could remember the sequence I should be able to complete it; fortunately I managed to do that!

The students were all from one or other of the fraternities, and they were competing to see which fraternity was the best. One fraternity invited us to join in their supper after *Ledersprung*, and we were very happy to accept; that was also another experience of a lifetime, when the students really let their hair down, with drinking, singing and much loud talking.

In England

At one time I was attached to Explosives & Chemical Products Ltd, a major explosives manufacturing plant in Alfreton,

Nottinghamshire, an area where many of the fictional events in the novels of D. H. Lawrence were supposed to have taken place. I was staying at Hole in the Wall Hotel; one evening I was having a drink in the bar, talking with a group of locals who, I discovered later, were influential farmers. They told me that there was a dilapidated cottage called Haggs Farm in the middle of an estate of the local squire, descendants of Jonathan Chambers; apparently D.H. Lawrence used to spend quite a lot of time at Haggs Farm, and he reputedly had an affair there with a daughter of the family, Jessie Chambers. Some say that Lawrence might have based "Lady Chatterley's Lover" on his experiences with Jessie.

Since it was based on a true story, the descendants of the original landowner did not allow anyone entering the grounds to see the cottage and, although it was heritage-listed, they were not only able to destroy it—some parts were demolished in the 1980s—but also to protect it from demolition. To discourage trespassers, it was patrolled by armed security guards, according to local rumour.

Being a great fan of D H Lawrence, I was very curious to visit Haggs Farm. The farmers decided that they would act as lookouts at strategic positions and let me know if they saw any security guards. They also told me the safest way to get there.

On the appointed day I followed the farmers' instructions, camera at the ready. I found that Haggs Farm was fully fenced off; at the gate was a large signboard saying something like: WARNING – Private Land. No admission. Armed guards patrolling. I heard one whistle, which was the "all clear". If heard three whistles, I was to get out of there, fast. After entering the ground it was easy to find the cottage. I went in and saw that everything was dusty but not falling apart. I took some photos then left. Once outside the gate, I gave a loud shout, "Okay!" My heart was thumping, and I was very happy indeed to reach the public road again. (Photo 55)

Photo 55: (top) Local farmers guarding my safety while I inspect the farm; (bottom) Haggs Farm where D.H. Lawrence wrote some of his books

Institute of Explosives Engineers

I mentioned earlier that I was involved with the Institute of Explosives Engineers (IExpE) in the UK. After arriving in Australia I started lobbying for the establishment of an Australian branch of the Institute. I found out that there were already a few members in Australia, mainly from the Defence sector, one of whom was a prominent member, Brigadier M. H. (Mac) MacKenzie-Orr, GM, OBE; however, there was no coordination or common activity between them.

In 1989, I organised an Explosives Forum at UNSW in order to inaugurate the establishment of the Australian Branch of IExpE. There were about thirty participants, and the branch was formally launched, with John Goold from the Defence sector as secretary and myself as Chairman.

Subsequently, several five-day blasting courses were organised under the umbrella of IExpE at UNSW, mainly targeted at blasting personnel who were practising in the industry, to refresh and update their knowledge. Some of the courses were followed by Explosives Forums to bring IExpE members together, with a formal dinner to close each Forum. (Photo 56)

UNIVERSITY OF NEW SOUTH WALES

GROUND VIBRATION AND AIRBLAST COURSE

3-5 December 1985

From left to right:
Marie Sykes (Unisearch), Peter Sciberras (Mines Inspectorate, N.T.),
Stewart Brown (Warkworth, N.S.W.), Martahan Silitonga (U.N.S.W.),
Brian Kennedy (I.C.I., Qld.), Ross Bennett (B.H.P., S.A.),
Mark Chambers (Drayton Coal, Qld.), Dick Benbow (Consultant, N.S.W.),
Frank Ford (Thies, Qld.), Vladimir Brusentsev (Pollution Control
Commission, N.S.W.), Dick Godson (B.M.I., N.S.W.), Eddie Brockhus
(Box Hill College, Vic.), Wally Malachek (Worsley Alumina, W.A.),
David Shearwood (DuPont, N.S.W.), John Goldburg (C.S.I.R.O., N.S.W.),
Alastair Torrance (B.H.P., N.S.W.), Greg Kennewell (Zinc Corp.,
Broken Hill), Stephen Thomas (North Mine, Broken Hill), Ray Cox
(Mines Inspectorate, S.A.), Gour Sen (U.N.S.W.), Mike Michaelsen
(Mines Inspectorate, Qld.). Missing in the photograph - Ron Connolly
(Drayton Coal, N.S.W.), John Hutchings (I.C.I., N.S.W.), Mick Kelly
(I.C.I. Vic.), Brian McClure (A.B.C., N.S.W.).

**Photo 56: Attendees of a Blasting related course organized by Gour
(standing far right)**

Unfortunately pressure of university work did not allow me to be further actively involved with the branch, and the Chair was taken over by Mac (now deceased). I was always available for consultation, however, and became the Patron of the branch. In 1993, IExpE head office awarded me with an Honorary Fellowship of the Institute. Since then I have continued to be interested in the activities and promotion of the branch. (Document 5)

In New Zealand

I was invited to give a course for practising blasting engineers in April, 1990, at the University of Auckland, titled "Explosives practice in quarries and mines: New developments". Some fifteen delegates from the mining and quarry industries attended. An integral part of the course was visiting sites to audit their blasting practices.

After the course was over, Brigid and I visited some tourist destinations, and we also met up with some friends and relatives living in New Zealand, including Brigid's cousin Noel and his family in Napier, and an ex-colleague from the 1970s Wynstones School days, Robin Bacchus, in Hastings. One of Robin's sons met us at the airport. He had a sticker displayed on his car carrying a slogan something like "NO MINING IN [SUCH-AND-SUCH] AREA". I wondered how he would react when he found out my background; and in fact, this did turn out to be a hot topic of conversation during our meal with the Bacchuses, where we discussed the pros and cons of mining versus non-mining.

Robin was heavily involved with the local Waldorf school, where his two boys were near the top of the school's curriculum. They were a delightful family; in their beautiful garden was a large tree loaded with figs, which I adore, but which none of the Bacchus family liked—so I had a whale of a time.

We also visited Stewart Island, off Invercargill at the southernmost tip of the South Island. At the time we visited

The Institute of Explosives Engineers

This is to Certify that

PROFESSOR GOUR C SEN PhD MSc BSc

was admitted as a
HONORARY FELLOW
IN RECOGNITION OF HIS SERVICES TO EXPLOSIVES ENGINEERING
AND TO THE AUSTRALIAN BRANCH

of
The Institute of Explosives Engineers on

17TH APRIL 1993

Given under the Seal of the Institute

President

Secretary

This Certificate is the Property of the Institute and is issued subject to the condition that in the event of Cesser of Membership it be returned to the Institute

Document 5: Gour's certificate of receiving Hon. Fellowship of the Institute of Explosives Engineers, UK

there were no hotels on Stewart Island, but there was a very pleasant bed and breakfast establishment where the owners treated their guests as family; at night we helped ourselves to drinks at the bar, and guests could go on fishing trips in the owners' large cruiser. I had never caught a single fish in my life before, but even I was catching thirty-centimetre ling every five minutes or so. I was amazed at what a good fisherman I was! But I was told later that the locals always know where the fish will be, and that everyone always catches some—which deflated my ego a little.

Soon after my appointment at UNSW, I was approached by a number of legal officers to give an expert opinion on various blasting-related disputes. Some cases required my appearance in court as an expert witness under cross-examination. In one case a miner had been injured in a blasting accident at Mount Isa Mines in Queensland, and he was suing the mine for negligence. Although there were explosives experts in Queensland, no-one wanted to take the case on for fear of antagonising the mine with possible retribution in the future. Hence the plaintiff's solicitor chose to approach me as someone who had no vested interest in the mine. I could see from the evidence that the mine had clearly violated its responsibility. My report was accepted and the presiding judge found that the case against the mine was proven, and the injured miner was awarded compensation. I heard the result from the lawyer, but he didn't provide many details.

****

In Japan

In 1991 I was awarded an exchange fellowship to Japan by the Australian Academy of Science. The primary purpose of the program was to support collaborative research between Australian and Japanese scientists. It was the first time this prestigious fellowship had been awarded to a mining engineer. (Document 6)

SCIENTIFIC EXCHANGES
WITH JAPAN - 1992/93

The Australian Academy of Science invites applications from Australian scientists who wish to participate in an exchange program with the Japan Society for the Promotion of Science between 1 July 1992 and 30 June 1993. Proposals in any field of natural science, basic and applied, including mathematics and engineering science, will be considered. The primary purpose of the program is to support collaborative research between Australian and Japanese scientists. Support will not be given when the primary purpose of the visit is to attend a conference. Applicants should be Australian citizens or permanent residents.

Short-Term Visits. Senior scientists may apply for short-term visits (3-6 weeks) to collaborate with Japanese scientists. A specific itinerary should be developed in consultation with scientists in the universities and institutions to be visited.

For short-term visits, the Academy provides an excursion international air fare and the Japan Society for the Promotion of Science provides an allowance for living and travel expenses in Japan.

Long-Term Visits. Long-term visits (6-12 months) to carry out collaborative research projects are also funded. Applicants should hold a Ph.D. degree or be about to submit a Ph.D thesis, in which case the visit would be conditional upon written confirmation that the thesis has been accepted for the award of the degree. Preference will be given to scientists who have less than five years of postdoctoral experience.

For long-term visits, the Academy provides an economy international air fare and the Japan Society for the Promotion of Science provides remuneration.

Letters of invitation from Japanese scientists to be visited must be submitted with applications. Participants may visit Japanese universities, National Inter-University Research Institutes or other research institutions affiliated with the Japan Society for the Promotion of Science. A list of institutes applicable to the program will be provided with the application form.

Proposals will be assessed on their scientific merit and on their potential for initiating collaboration and expanding contacts between Australian and Japanese scientists.

Applicants are asked to state why the work they propose to do would best be done in Japan and to discuss how the work would benefit Australia.

The exchange program is funded by the Australian Government.

Application forms and a list of institutes are available from:
International Exchanges
Australian Academy of Science
GPO Box 783
Canberra, ACT 2601
Telephone enquiries: (06) 247 3966, Bonnie Bauld or Judith Hlubucek.

**Document 6: Recipient of Japan-Australia Scientific Exchange
Scholarship**

Document 6: *Contd...*

Australian Academy of Science

Ian Potter House, Edinburgh Avenue, Canberra 2601

RI 24
17 October 1991

Phone:
(06) 247 3966

Dr. G.C. Sen
Department of Mining Engineering
University of New South Wales
PO Box 1
Kensington, NSW 2033

Dear Dr. Sen,

I am pleased to inform you that your application to participate in the Academy's exchange program with the Japan Society for the Promotion of Science was successful. The Academy's Japan Exchange Committee approved your proposal to visit Japan for 28 days during the 1992–93 financial year. JSPS must now approve our Committee's nominations, but no problems are expected.

Under the terms of our agreement, the Academy will pay your travel agent for a return excursion air fare from your home city to Tokyo, a departure tax stamp and ticket cancellation insurance. JSPS will provide you with living and travelling allowances and medical insurance for the time you spend in Japan. All costs for persons who may travel with you are your personal responsibility.

After the Academy has received final approval for your visit from JSPS, I will send you a travel warrant to be given to your travel agent when you book your airline ticket. You should provide me with details of your travel arrangements at least six weeks prior to your departure, so I can officially advise my counterpart at JSPS. If the timing of your visit changes for any reason, please advise me as soon as possible.

It is a condition of acceptance of the award that you submit to the Academy three copies of a short report of your visit within 30 days of your return from Japan. Guidelines for the report are attached.

Congratulations on the success of your application. Please confirm in writing that you expect to be able to accept the award and travel to Japan during the 1992–93 financial year.

With best wishes.

Yours sincerely,

Bonnie Bauld

Bonnie Bauld
International Exchanges Officer

In due course I travelled to Japan and was attached to the Mining Department at Hokkaido University (Professor Nakajima). I was interactively involved with their research on rock mechanics in Sapporo. This work has particular relevance to the stability of the rock in underground excavations.

The most memorable event was my visit to the Toyoha base-metal underground mine near Sapporo. Underground, the temperature of the rock was close to 100°C, making the air too hot for the miners to work in without some cooling arrangement. The working area was air conditioned by a network of pipes circulating crushed ice, and the miners rested from time to time in an air-conditioned tent near the work site; a cold water spray at the rock face was also used for cooling. The conditions underground were so oppressive that I could not tolerate the heat for more than half an hour, and had to leave. I understand that the mine is now closed. (Photo 57)

Photo 57: Gour in a very hot underground copper/gold mine in Hokkaido, Japan

After my return to Sydney, Professor Nakajima came to UNSW as a Visiting Professor, but soon after I had briefed him on the proposed collaborative research project he fell ill and had to be hospitalised. When his health had somewhat improved he was advised to return to Japan, unfortunately ending this opportunity for collaboration.

18

Third Sabbatical

Apart from some technical visits in India and presenting lectures at various institutions, I spent my third sabbatical leave, December 1991 to July 1992, mainly in Sweden and Canada.

In India

After visiting a very deep underground gold mine (over three thousand metres) in Kolar, Brigid and I went to Trivendrum to see our second Plan International- sponsored child. It was another memorable occasion being welcomed into the child's village by an ensemble playing drums and *shehnai* (an Indian wind instrument, rather like an oboe or clarinet). (Photo 58)

Photo 58: (top) musical reception of our visit; (bottom – arrows) our World Vision sponsored child, India

We saw at first hand how the sponsorship money not only supported the child but a significant amount went towards the welfare of the village.

In Sweden

I started this part of my leave at the end of the Australian academic year in December 1991, and spent most of time in Luleå in the far north of Sweden, near the Arctic Circle—the latitude of the "midnight sun" where the sun never sets in summer and never rises in winter. It was a pleasure to work at the University of Technology there; one significant research program of interest to me was an investigation into blast-induced fragmentation influenced by the joints in the rock.

Naturally, it was very cold in December—between –6° and –20°C. The university provided us with a small, centrally heated, cosy house in the centre of the small city. (Photo 59)

Photo 59: Our home for six months in Lulea, Sweden

I went the five or so kilometres to the university by bus every morning; they had offered me a car but I didn't feel confident driving on the icy roads.

Soon after we got there we realised that our winter

clothing wasn't going to be adequate for northern Sweden, and kitted ourselves out with well-insulated shoes, coats, hats, scarves, gloves and so on to be comfortable walking about outside. When we first arrived in early January there were about two hours of daylight; the sun rose in one corner of the clear sky at about eleven in the morning, made a small arc and dipped again to set soon afterwards

Luleå is on the Lule River, which was frozen almost solid; navigation wasn't possible in winter, and it was interesting to see people ice fishing through holes they had cut in the ice.

One weekend we went by bus to Kalix, about fifty kilometres north-east of Luleå, and stayed in a holiday cabin let by a very hospitable farmer who gave us our first experience of riding a snowmobile through the pine forest. (Photo 60)

**Photo 60: Our experience of 'Snow mobile' transportation
in Kalix, North Sweden**

We also had a meal of wild boar meat there. One clear night we saw the aurora borealis—the northern lights—a fabulous display of patterns of coloured light dancing across the sky. Brigid went outside to experience this wonderful natural phenomenon in the open—it was too cold for me, though; I was content to watch it through the window. We also tried out some borrowed skis while we were there, but about half an hour and several falls later we realised that we were getting too old to take up skiing. I published our experience of Kalix on the university blog. (Photo 61)

A memorable short-break

One weekend my wife and I had a happy experience which we should like to share with others, in particular with visitors from other countries.

We always like the outdoors and natural lifestyles, so whenever we get an opportunity to get away from the city, we take it. While browsing through brouchures from the local tourist office in Luleå we found one on the so-called "Vilt-farm" at Marieberg about 20 km from Kalix. This is a small farm where they rear dear and wild boar, and have small cottages for visitors, open all year round.

We travelled by bus to Kalix where Eva Enström, the hostess, very kindly came to collect us in her car. We arrived at the farm on Saturday 21st March, at about midday. The accomodation in the hut is really good, clean and cosy. Everything you need is provided except for food which we were asked to bring as they are self catering huts.

Thrilling experience

Soon after our arrival the activities started — seeing the animals fed and an exciting ride on an old army truck into the woods to see some of the surrounding landscape which is really enchanting. The owners very kindly lent us a "spark" and even their skis, and we practised on these almost till sunset since that was our first encounter with these two "toys". What a thrilling experience!

Then we used the sauna and cooked ourselves a meal. The evening was crispy and freezing, and the elements of the nature were ideal to see a spectacular display of the Northern Light in that dark surrounding.

The following morning we had a ride on the "Snow Mobile" along the frozen Kalix river and through some snow-covered tracks in the woods. Later we went for a walk through the woodlands where we found and ate bunches of ripe lingonberries. At about 4 o'clock we were driven back to Kalix to get the bus for Luleå.

This short break was a wonderful experience of a rural part of Sweden, about one hour's drive from Luleå. It´s natural beauty and hearty hospitlity. We hope that this account will inspire others try it out for themselves.

Gour Sen, a guest researcher at Mining and Rock Excavation division

Photo 61: Our experience of week-end in Kalix as published in the Lulea University Magazine

On another occasion we went to the annual Lapplander fair in Jokkmokk, just north of the Arctic Circle. Seeing the array of local Lapp produce and the extravaganza of colourful garments and utility products was another wonderful experience. (Photo 62)

Photo 62: (top) Lapps' annual festival beyond North of Arctic Circle which is shown (below) by an arrow

In Luleå one evening we went to see a modern version (in Swedish) of Shakespeare's "As you like it" at a local theatre. It was a hilarious performance, and we enjoyed it so much that the language difference didn't matter; afterwards, as we were leaving the theatre, each man in the audience was given a packet. We thought it must have been chocolates for us, so we opened it at home to have with our coffee—some surprise when we found it was a condom!

There was an accident in the house where we were staying, which I include here as a cautionary tale: Brigid was boiling a saucepan of water for a cup of tea, when the phone rang. She answered it and became immersed in a long conversation. Suddenly remembering the boiling water, she rushed to the kitchen to find that the water had boiled away and the saucepan was red hot. She grabbed the saucepan off the stove but the metal had been burnt through by then—the bottom of the saucepan, red hot, fell and burnt a hole in the lino floor covering. (We were not very popular when we told the university people about it.)

In Russia

On the recommendation of Professor Gunnar Almgren of Luleå University, Brigid and I visited the Leningrad Mining Institute in Russia. We arrived at Saint Petersburg railway station late one afternoon in early April, 1992. The station building was very poorly lit, and there was no kind of information booth we could locate. We took a taxi to Hotel Moscow where we had reserved a room for four nights. The scarcity of food at that time was quite evident from the menu in the restaurants. Hotel breakfast was very frugal—a kind of porridge, two pieces of toast and a cup coffee. No use asking for anything more! A young man sitting next to me could speak English sufficiently well that we could have a conversation. I discovered that he was from Siberia and his business was to sell imported poker machines in Russia. In the course of conversation he told me that he knew where to buy good quality genuine caviar and vodka, so one day I gave him the money and, while we waited outside the shop, he bought them for us. (Photo 63)

The next day we went to the Mining Institute and met Professor Boris Abramovich who showed us around. I was astonished by the vast geology museum at the Institute. I

**Photo 63: A Russian (shown in the centre) helped us
buying genuine Vodka and Caviar, St Petersburg**

had never seen such a huge collection of beautiful minerals
in my life. Professor Abramovich invited us to dinner in his
apartment, where we were pleasantly surprised to see very
many special food items on the table. It appeared that one
had to have connections to obtain items that you could not
buy in the shops. His family and a number of his colleagues
were there as well, and there was no shortage of lavish quality
and quantity on the table.

Abramovich introduced us to his daughter and her
boyfriend. They both spoke English, and later they took us
sightseeing to such places as the State Hermitage Museum
(formerly home of the tsars) and the Summer Palace of Peter
the Great.

In Canada

Here I was attached to Queen's University in Kingston,
Ontario, and I participated in a research project on the
correlation between the intensity of ground vibration and
the in-hole velocity of blast detonation. In our free time at
weekends we visited Montréal and Niagara Falls. In Québec
we indulged ourselves by staying at the Fairmont Le Château

Frontenac, which was a wonderful experience. We were pleasantly but rather surprised to see a version of Nelson's Column in Montréal in 1809. Its existence there still raises controversy in Québec, which was a French colony long before Canada became part of the British Empire. (Photo 64)

Photo 64: Nelson's statue in the centre of Quebec , a French dominated city in Canada!

I hired a car in Kingston, which was my first experience of driving on the right-hand side of the road. There were a couple of nail-biting moments when I took the wrong entrance at road junctions and found with horror other cars coming from the opposite direction.

A couple of weeks of after we arrived in Kingston the frozen Lake Ontario was gradually thawing out, so I joined the Kingston Rowing Club. It was a great pleasure to meet so many friendly rowers there. Apart from my usual sculling I often joined in Fours or Eights, and even participated in a regatta. I still have the singlet with Kingston Rowing Club's logo that was presented to me after the regatta.

19

Promotion and the Co-op Program

A fter returning to Sydney I thought that I now had enough credentials to apply for promotion from Senior Lecturer to Associate Professor. Again, an enormous amount of material was needed to make a portfolio to accompany the application, which was then subject to peer review, followed by an interview where I was grilled by questioners; I must have impressed them, and gained the promotion at the end of 1992.

In the early 1990s a new initiative called the School of Mines Co-op Program was formulated. The main thrust of this venture was to attract high-calibre students into the mining industry by offering them substantial scholarships for three years (at the beginning it was $10,000 each year) and providing paid practical training during the university's long vacations. A panel of three—two from the mining industry and one from the university—interviewed school leavers applying for the scholarship. At the interviews, factors such as the candidate's motivation, academic performance and extra-mural activities were judged and scored. In a typical year, scholarships were offered to six of about a hundred applicants. After culling, some twenty applicants were selected for interview. I was the coordinator of the program from not long after its commencement till my retirement in 1996.

In order to obtain scholarship funding, I spoke with mining executives and visited a number of mines. If the mines were in remote locations, I asked Brigid to accompany me on the long drive. There were three incidents worth mentioning.

The first occurred on our way to the Pilbara from Perth in Western Australia. I hired a car in Perth then started out for Mount Newman one morning. The road was dead straight, and we could have counted on our fingers the number of cars we passed going the other way. After a couple of short breaks on the way, after eight hundred kilometres we reached Meekatharra and stayed overnight there at a motel. From there, Mount Newman was about four hundred kilometres further on. The next morning I topped up the petrol tank, and wondered if I should carry some spare petrol, say a ten-litre can. At the service station I was told that there was another one about half-way to Mount Newman. With that assurance we set off, and we eventually came to the service station.

As soon as we stopped, a man came out of a hut and shouted, "No fuel, mate!"

"So what do we do?"

He replied that he was expecting a delivery next morning, and we could stay the night in one of the cabins nearby. That wasn't possible, as I had a very important meeting early the next morning. The only option was to take a chance, and drive at no more than sixty kilometres an hour and turn the air conditioning off (although it was the middle of summer). We knew we would be stranded on the road in the heat of summer if the car ran out of fuel, and would be unlikely to see a passing car to assist us. In spite of that possible danger, I decided to take the risk and drive on. We were tense the entire way, and not a word was spoken. At last we reached Mount Newman, where a service station lay a few metres ahead. What a relief—there were less than two litres left in the tank.

Another incident occurred on a trip to Ranger Uranium Mine in the Kakadu National Park, Northern Territory, which involved driving some two hundred and thirty kilometres from Darwin. Michael Lindsay, one of my former students who had worked at the mine some years earlier, had suggested a considerably shorter route via an unsealed road. After arriving in Darwin airport there was an unusually long delay in arranging to pick up the rental car, and by the time we left Darwin it was getting dark. We debated whether we should take the longer, sealed road or risk the shorter, unsealed road. We would arrive at the motel very late if we took the longer route, so we started to follow the unsealed road.

After driving more than an hour we had not seen another vehicle going either way. Then we came across a little creek crossing the road, with a signpost saying DANGER! CROCODILES! What should we do? Go back to the sealed road? This incident was well before the mobile phone age, of course, so there was no way of communicating with the motel. I tried to see how deep the creek was by dropping a stone in it, it was just swallowed by the water. Then, not to waste any more time I decided to drive through the creek anyway, thinking that if we were stranded mid-stream we would have to climb on top of the car beyond the reach of the crocodile. My heart was thumping while I slowly drove through the water, which at one point was lapping over the doorsill and starting to seep into the car—but shortly the front of the car tilted up, and to our relief we were in dry land.

The third incident, for a change, had no danger element. I had learnt from experience that, if it were necessary to get in touch with someone in a key position, it was better to phone myself rather than arranging it through my secretary. On this occasion I needed to arrange a meeting with Jerry Ellis, the chief executive of the major mining company BHP. Invariably his secretary answered my calls, one of her jobs of course being to filter incoming calls. When I asked her for an appointment with the CEO, her immediate answer was that his diary was full. When I said that this call was coming from the Head of the Mining Department at UNSW, she somehow found a spare half-hour and made the appointment.

When we met, he told me that I was the first academic to have a meeting with him. I said, "I'm not surprised. I'm sure many academics have tried to contact you, but your secretary would have shielded you from those sorts of meetings." Anyway, my mission to arrange scholarships was successful; a spin-off from the meeting was that he inspired me to write a book on blasting that could be carried by his engineers in their pockets at the mine sites. *Blasting technology for mining and civil engineers* was published by UNSW Press in 1995. (Photo 65)

Photo 65: This was first book published by the University of New South Wales Press

I was pleasantly surprised to be awarded an Honorary Fellowship by the UK-based Institute of Explosives Engineers. In the certificate of commendation it reads: "In recognition of his services to explosives engineering and to the Australian branch".

****

The Head of the mining Engineering Department, Dr E. G. (Ed) Thomas, took early retirement in 1991 for health reasons, and I was appointed Acting Head; subsequently in 1992 I took on the position of the Head for three years. During this period I offloaded some of my duties concerning the Co-op Program and appointed a Senior Lecturer, Dr Chris Daly, to look after that area.

There were various strategies at the back of my mind, and as soon as I was in a position to do something about them, I began to implement them one after the other. The first was to have a short (half-hour) staff meeting every fortnight to bring everyone up to date on what was happening in the Department. Brief minutes of the meetings were taken and circulated to staff. These meetings were held in the small room I set aside as a place for the staff to meet in their breaks, which also acted as place to meet for discussions with visiting dignitaries and personnel from the industry. This was another of my innovations for the Department.

The third one was to create a strong alumni culture. In order to establish the foundation of such a culture I arranged for a group photograph to be taken of the final-year students with staff members every year. Fortunately the school office had a list of past students' names and addresses, although some were out of date. I then organised an Alumni Forum with a dinner, to which all known alumni members were invited. An alumnus by that time holding a key position in the large mining company Newmont Mining Corporation agreed to deliver the keynote speech. It took nearly two years to organise the event, but it was a great success. My vision was to hold similar gatherings every three years or so, but the idea was not followed up after I retired.

In my early days at the school I observed that many students completed the course without having adequate practical mining experience. When I took over the headship my forth strategy was to address the problem, which I thought could be solved by arranging training positions for all first-year students at the start of their summer vacation. For this I would need the cooperation of the mining companies, which would not usually be forthcoming by sending letters or by telephoning. I needed to see some of the key personnel of various mining companies, which I started to do. Sometimes they were dubious about offering paid positions to students, objecting that, in effect, they wouldn't get value for their dollar if they employed inexperienced students. My prepared response to this was, "This experience has to start somewhere. You can put a student in a very unskilled job, such as cleaning up spillage at conveyor belt transfer points. If they were involved in a real mining situation they would surely learn some of the activities of mining operation." I obtained offers sufficient for every first-year student to have a mining-related job at the end of his or her first year, and I used to tell them that they should impress the officials with their work so that they would get recommendations for their future employment.

During every university vacation I travelled to various mines all over Australia to audit the students' performance. This scheme worked so well that I didn't have to find jobs for the students in other years; however, the practice wasn't pursued on that scale after my departure.

****

China Assignment

One day I had a phone call from the Chinese Consulate asking if I would accept an academic, Mr Dexin Ding, for one year. He was interested in acquainting himself with my research area. I agreed, provided all his expenses were borne by his government. When he arrived I introduced him to Frank Santoro, one of my postgraduate students, who would in turn get help for his fieldwork from Mr Ding, so that the arrangement would accelerate Frank's progress, too.

A couple of weeks later I saw Frank and asked him how his work was progressing. He said, "No change. I haven't seen Mr Ding since I first met him."

At that time there was a large contingent of Chinese students, and it wasn't difficult to track down Mr Ding. When I asked him what he had been doing for the past fortnight, his smart answer was, "I was in the library, familiarising myself with the background of the research project."

This was the standard answer that no-one could verify or refute; however, I heard through the grapevine that he was working at a supermarket, filling up shelves. Although I was angry with him, I didn't want to put him into a corner, so I said, "When your government asks me for a report on your progress in a year's time, my answer will be f... all."

"What does that mean?" he said.

I said, "Ask your mates, they'll tell you. In the meantime, you report to Frank every day and help him, and I am expecting his report in a couple of weeks' time."

After that Mr Ding's work was so good that we produced a joint paper for a technical journal. When the year of Mr Ding's stay was over he wanted to stay another year, working on my research projects. He then asked me to strongly support his application to his government for the renewal of his scholarship. I followed this up, and he was permitted by the Chinese Government to continue working with me; subsequently his progress was very good.

Shortly before his departure date he came to see me looking most distraught. He told me that his bank in Australia would not allow him to send the twenty-four thousand dollars he had saved from his earnings back to China. I went to talk to the manager of the campus branch of the State Bank (now part of the Commonwealth Bank of Australia), John Weston, whom I knew very well. At that time the government had, quite rightly, placed restrictions on foreigners moving large sums of money to discourage money laundering connected with drug trafficking and other criminal activity. I explained to the manager that it was possible for a foreign student to earn and save this amount by working the permitted twenty hours a week, and I certified that the money was genuinely earned and that he should be allowed to send it; so the episode was happily ended.

As it happened I took the department's headship at the beginning of an austerity period imposed on all Australian universities by a series of cutbacks in government funding. The next cutback was to reduce spending in all areas of the department. It wasn't too difficult for the first year, but a ten per cent reduction in recurring expenses had to be implemented, so I decided not to extend my headship after the end of my term. Professor Jim Galvin, who had joined the staff as a researcher from the mining industry not long before this, was likely to be my successor. However, the Dean of the Faculty of Applied Science, Professor Gerry Govett, was unsure of the wisdom of giving this responsibility to a person with relatively little academic experience. Govett asked me if I would consider continuing in the headship for another year, but by that time I had made up my mind to quit, so I declined; moreover I had already arranged another sabbatical leave starting at the end of the first semester, 1995.

****

Organising Students Technical Visit to Japan

During my recent visits to Japan I realised that the students would benefit from practical exposure to mining in an overseas country. When I approached the senior students with this idea they were more than enthusiastic, but cost was a problem, considering the high cost of living in Japan.

I had very good rapport with Professor Kenichi Uchino at Kyushu University, who agreed to be the coordinator in Japan and make the travel and accommodation arrangements there for our group, but cost was still a big obstacle for the students.

In order to make this trip affordable, I asked a number of Japanese companies involved with the Australian mining industry if they would be willing to sponsor such a project. Their support was very encouraging, offering sponsorships that covered air fares, internal train fares, accommodation and most meals for eight days, and the students to pay only $1000 each for the trip.

Brigid and I took a group of seventeen students to Japan during the university's Easter break in 1994 for eight days. After a couple of days in Tokyo to see the razzmatazz of the great metropolis we travelled to Kamioka where we met Professor Uchino and his daughter.

We stayed at the Kamioka Mines guest house. As soon as we arrived there all our students went outside to enjoy the snow; they were like young children, running about and throwing snowballs. The next morning we visited the Mozumi Mine where a neutrino observatory was located underground. The experiments there contributed substantially to the advancement of particle physics, although it did not achieve everything it set out to do and has since been superseded.

On another occasion we visited a surface limestone quarry on Shikoku Island where the students were impressed by the remote-controlled maneuvering of loaded dump trucks that travelled to the unloading point, unloaded and returned for loading again. After visiting Miike Colliery, which was an underground coal mine (now closed), our students joined local students at a social function. (Photo 66)

This was the first-ever UNSW overseas technical visit by mining engineering students, and it was certainly a success.

Photo 66: A social function organized by the Kushu University students for UNSW students, Japan

20

Fourth Sabbatical

My final SSP leave from UNSW before my retirement was spent in England, Poland, Czech Republic, Germany, Brazil, South Africa and Zimbabwe, between July, 1995 and February, 1996.

In England

Here I toured round for about ten weeks, to familiarise myself with the activities of an explosives manufacturing company (Exchem Explosives Ltd), a consulting firm for blast-induced vibrations (Vibrorock) and the Camborne School of Mines in Cornwall, and their contributions to the mining industry.

In Poland

I was in Poland to present a paper at the 26th International Conference of Safety in Mines Research Institutes held in Katowice. While I was there I took this opportunity to visit Kleofas coal mine where a coal seam of four metres thick was extracted in one sweep, which required a clever technical procedure.

The conference dinner was held in an underground rock salt mine at Wieliczka, not far from Kraków, Poland's second largest city. Mining began there in the thirteenth century and only ended in 1996. While commercial mining was still carrying on, workers carved fantastic sculptures, including a complete chapel, from the solid rock salt. These were so beautiful that the mine is now a UNESCO world heritage site, and it is a very popular tourist attraction. (Photo 67)

Photo 67: Sculptures created from solid rock salt in a mine near Cracow, Poland

Brigid and I spent a couple of nights in the small town of Prszczyna to experience something of the Polish countryside. After checking in our hotel we were taking a leisurely walk, when we heard music coming from a loudspeaker some distance away; we found a large crowd gathered in a field, with apparently some kind of fair or festivity in progress, but it wasn't clear to us what was being celebrated. The language foiled us for a while, but eventually Brigid, who speaks German, was able to ask someone about it. They told us that it was the local electricity company's annual function. Among the many stalls, an old military tank, converted into a kitchen, was serving goulash. We were not really hungry, but it smelt so good that we had to try it—delicious!—better than you would get in most restaurants. (Photo 68)

Photo 68: An old army tank converted to make 'goulash', Poland

Then we went by train to Warsaw to catch our flight to London. At the station, as we were trying to get on the airport bus with our suitcases along with a crowd of people trying to do the same, I discovered that my wallet—containing money, driver's licence and credit cards—was missing. No wallet! My pocket had been picked! Brigid had some Polish zlotys to pay for our bus tickets, luckily, and we had plenty of time before the flight, so I reported the theft to the police station and it was duly noted in their logbook, but they didn't think there was much hope of recovering the wallet. In London I reported the theft to Visa and American Express, and thankfully they each promptly gave me a replacement card.

Postscript: In spite of reporting the loss and receiving a replacement card, the thief used the stolen Amex card for four months to buy some expensive items. Before I had left Sydney for overseas I had arranged direct debit payments on my credit card, so I didn't know that total fraudulent purchases had come to more than four thousand dollars; it was the responsibility of Amex, as the credit supplier, to prevent fraudulent transactions, and the money was eventually reimbursed.

In Czech Republic

Here the Vice-Rector, Prof. Ing. Václav Roubícek—who has since died—gave me a comprehensive tour of the Technical University of Ostrava (the third-largest city after Prague and Brno), the only institution in the Czech Republic where a course in mining engineering was offered. The geology section was particularly impressive. We were accommodated during our stay in a well-appointed—in fact rather posh—apartment owned by the university; later we found out that in Soviet days it had been used only by top visiting Communist officials.

On the following day I went to visit the Lazy Mine underground colliery in Orlova. The mine had a spontaneous combustion problem in the worked-out area (not unlike that in the Northern Coal Field of New South Wales), and the engineers had developed a novel method to control the hazard.

We then went to the capital, Prague, and there spent a few days in a small hotel, sightseeing and taking in the city, and meeting up with the Thompsons, old friends from England.

In our hotel there was a bar, which was tended by the owner during the day. He handed over to the man responsible for the bar at night. Before leaving the hotel the owner marked the level of liquor in each bottle; on one occasion we saw the night-shift man help himself to some vodka, then top-up the bottle by adding water.

In Germany

Here we were the guests of the Clausthal University of Technology, mainly for me to attend a seminar on the rehabilitation and the use of space in worked-out mines.

I also made technical visits to a quarry in Bergwerksanlage producing basalt rock for the construction industry, and to the explosives manufacturer MSW-Chemie GmbH in Langelsheim.

In Brazil

I was attached to the Mining Engineering Department at the Federal University of Rio Grande do Sul in Porto Alegre, which is in southern Brazil where most of population were descended from European immigrants. We had an apartment in a boutique hotel within walking distance of the university. People here were extremely friendly, in particular Professor Jair Koppe, the head of the mining department—in fact, so much so that social activities took up a large part of our free time there. Brazil is the only country in South America where Portuguese is spoken, so the language was a bit of a problem for me; however Brigid used her linguistic skills to overcome the difficulties most of the time. Whenever I had to do any professional auditing work in mines a university staff member, Jorge Segui—who has since migrated to Australia and works in the mining industry—accompanied me as interpreter. He also drove us to various places for my work-related trips. When Jorge drove, I would sit next to him with my eyes shut most of the time, so frightening was his adventurous driving.

While travelling, we often saw wild orchids growing by the roadside. Since I am an orchid fancier, this experience was really thrilling. Another new experience was to have meals in the *churrascaria*, Brazilian steakhouses where churrasco (roughly meaning barbecued in Portuguese) meat, fish and vegetables were served at the table from skewers;

the food was served until the diners could eat no more. Some large *churrascaria* also provided an acrobatic floorshow.

On one occasion, after visiting an open cut coal mine, we went to Rivera, a city on the border between Brazil and Uruguay. In 1943, the Plaza International was built there straddling the border in celebration of this peaceful relationship, and is perhaps the only international square in the world. The border community was unusual in that people from both Livramento (Brazil) and Rivera (Uruguay) were free to move anywhere throughout the twin-city community. It was easy to get lost in the suburbs and not know whether you were in Brazil or Uruguay. Customs and border checkpoints were located outside the city. A plaque in the centre of a street marked the nominal border; on the plaque

**Photo 69: In Rivera the border between Brazil and Uruguay
is shown by an arrow on a plaque**

a curved line and two arrows pointing in opposite sides indicated the directions of Brazil and Uruguay. (I also remember that cigarettes were cheaper in shops on the Uruguayan side.) (Photo 69)

After my official stint at the university in Porto Alegre we went on an extensive sightseeing tour of Brazil, using air passes we had bought in Australia for the various Brazilian sectors at quite reasonable prices. In that way we visited Iguazu Falls near the border with Argentina, Amazon rain-forest, and some northern and north-eastern cities in Brazil. (Photo 70)

Photo 70: Brigid and Gour enjoying Iguazu Falls, Brazil

On one occasion we stayed in a cabin deep in the Amazon rain forest for three nights. Because there was no road, we were taken there in a small boat with an outboard motor, following a water course. The meals were very frugal but while we were there we saw beautiful wild orchids, birds and reptiles. Despite many warnings, we had not been vaccinated against malaria, dengue fever, yellow fever or any other tropical diseases. Our strategy in the Amazon rain forest was to avoid going out in the evening or early morning, and to apply anti-mosquito lotion liberally on every exposed area, and luckily this protected us from being infected.

In South Africa

This second visit to South Africa was prompted by a request from Gencor, the large South African mining company, to

PROGRAMME AND ITINERARY - Prof G C Sen

(3 to 26 January 1996)

Visits and Consultations with Samancor, Gengold and Ingwe - Divisions of Gencor Ltd

Day	Date	Activity	Notes	Resp
Wed	3	• Arrive Johannesburg flight RG 828 at 12:10. • Transport to local hotel • Rest of day at leisure	Met by PJH Short	PJHS PJHS
Thurs	4	• 09:00 met by PJHS for visit to Samancor Mineral Mines	Pretoria Lyttelton Dolomite Mine - urban environment - ground vibration - air blast - dust amelioration - final wall trimming Delmas Silica Mine - - dust amelioration - fragmentation optimisation	PJHS/JRVC
Fri	5	Samancor Mineral Mines		JRVC
Sat	6	At Leisure		
Sun	7	Picnic/Game reserve visit	PJHS & family	PJHS
Mon	8	Samancor Mineral Mines		JRVC
Tues	9	Samancor Mineral Mines		JRVC
Wed	10	• 06:25 flight SA 1145 Johannesburg to Kimberley ETA 07:45 • Transport to Hotazel • per local programme	• Samancor transport • Mn mines transport	PJHS CFN
Thurs	11	Mamatwan mine (opencast)	Programme for Mrs Sen also arranged	
Fri	12	Wessels mine (underground)		
Sat	13	• Game farm visit • Transport to Kimberley • Flight SA 1068 Kimberley to Johannesburg ETA 14:30 • Transport to local hotel	Mn mines transport Samancor transport	CFN PJHS
Sun	14	At Leisure ...Lunch with the Salamon family		

Document 7: Gour's consultancy programme for Gengold requiring auditing a number of mines situated wide areas in South Africa

Document 7: *Contd...*

Mon	15	09:00 collected by Gengold...travel to Welkom Local introductions in afternoon		PDKR
Tues	16	Orientation and visit to St Helena Gold Mine		
Wed	17	Visit to Beatrix Gold Mine		
Thurs	18	Visit to Oryx Gold Mine		
Fri	19	Open forum discussions and feedback meetings - Gengold managers (pm return to Johannesburg as alternative to the next day at own choice)		
Sat	20	am Return to Johannesburg local hotel		
Sun	21	At leisure		
Mon	22	07:00 collected by Optimum Colliery staff. Drive to Middelburg area. Field visits to drilling & blasting operations Hotel = Midway Inn, Middelburg		HCW
Tues	23	Blast monitoring equipment and practices		
Wed	24	Bulk explosives site and explosives loading practices		
Thurs	25	• Visit Middelburg Mine Services and Rietspruit collieries • Return to Johannesburg hotel in afternoon		
Fri	26	08:00 flight to Zimbabwe	Samancor transport	PJHS

PJHS = P J H Short   tel 011-491 7324   768 2901 (h)   082-553 9602 cell
JRVC = J R V Caddy   tel 012-664 5649   47 7266 (h)   082-552-2170 cell
CFN = C F Nel   tel 05374 260
PDKR = P D K Robinson tel 011-376 3489   476 3778 (h)
HCW = H C White   tel 0132 96 5111

Local Hotel = The Courtyard, Rosebank
               cr Oxford & Tyrwhitt ave
               opp Firs Shopping Centre
               Rosebank   tel 011-880 2989
                                 fax 011-880 8425

audit their blasting operations at a number of their mines. The schedule of mine visits was exhaustive and required a lot of travelling. (Document 7)

There had been significant political changes in South Africa between our first visit in 1984 and this one (1996). Of course, by now the apartheid policy had disappeared, and the country had been under the presidency of Nelson Mandela since 1994; however, killings, kidnappings and highjackings were still plaguing the country. We were put up in a hotel in Sandton, a Johannesburg suburb heavily protected by security guards, and were advised not to leave there unless in the company's car and with the company's driver.

We were driven to Johannesburg airport every Monday morning to travel to different mines, and returned to the hotel at weekends, which were rather boring because we were so isolated. Sandton had many very up-market and expensive shops and restaurants, yet there was always a pervading feeling of living an artificial life inside a barricaded zone. We loved being in the mining towns during the week where there was no violence to anything like the extent that was happening daily in Johannesburg.

In one of these trips we visited the worked-out Kimberley diamond mine pipe in Northern Cape Province, where mining began in 1867 and finished in 1914. At two hundred and forty metres it is the world's deepest mined open hole. (Photo 71)

We also drove to Hotazel, to visit an open cut manganese mine in the Kalahari manganese field, also in Northern Cape; the manganese ore grade was very high, averaging sixty to seventy per cent. Hotazel was originally called Hot As Hell but, so the story goes, the local priest didn't like the name, and changed it to its present spelling. It is still pronounced Hot as Hell, though.

We stayed at the mine's guest house for a couple of nights. The arrangement was that at dinner time we would be hosted

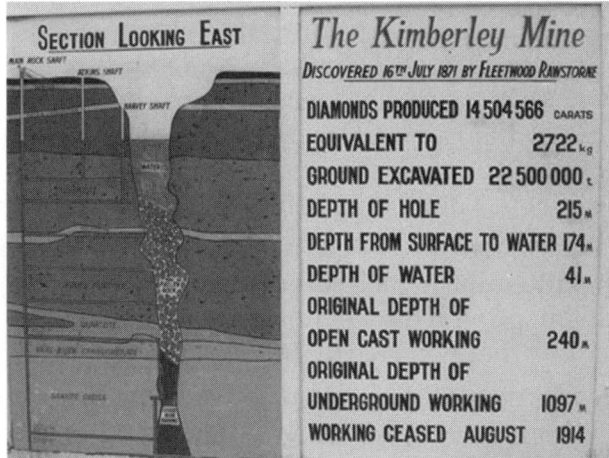

Photo 71: Statistics of Kimberley Diamond Mine, South Africa

by one or other senior mine official. One evening we were told that no senior official was free, so we were left on our own. Then we heard that the General Manager, Mr C. F. Nel, was going to join us. We guessed that the GM must have felt awkward about us being left on our own, and that his visit would be no more than a duty call. Nel was a big man with a walk like a lion's. At first he was very formal and rather cool, then he asked if we would like to go to the mine club for a drink before dinner. We could feel the aura of his authority—whenever we passed anyone there was a definite sense that being with the GM was pretty much the same as being with God.

In the club bar he said to Brigid, "I'm having beer, what's yours?"

"Same as yours."

That response seemed to have a huge impact on him. Suddenly the officious General Manager became a friendly soul. We started swapping jokes, and the evening became a very pleasant social *tête à tête*. In fact, when we went to say goodbye the next morning he presented us with a very handsome book on South African wildlife.

In Zimbabwe

I had two reasons to visit Zimbabwe—to audit a platinum mine in Chegutu (formerly Hartley) about eighty kilometres south-west of the capital Harare, and also to visit a New Zealand mining academic's father, Bill Rainsford, in Shurugwe, about a hundred and fifty kilometres from Harare.

It took me about two weeks to audit the blasting practice of the Chegutu mine, which was owned by BHP at that time. The company provided me with a car and a house for us in Harare, from where I drove to the mine each morning.

There was no fear of being robbed or being kidnapped in Harare in those day, and we could go shopping or into town without fear. For health reasons we were told not to hang towels on an outside line because of the torsalo fly. If it laid eggs, they would hatch and the parasitic larvae penetrate the skin and lay more eggs, which would produce blisters and worse health problems.

Like many tourists we also visited the strange ruins at Great Zimbabwe, and Victoria Falls where we crossed the bridge over the Zambezi River, which is the border between Zimbabwe and Zambia, as well as various local places such as Bulawayo. (Photo 72)

After that we went to stay a few nights in Shurugwe with the Rainsfords. Bill owned the small Pompeii gold mine, which was on their property in a rather wild and remote part of central Zimbabwe. Only a couple of miners were employed but they produced enough gold to justify the mine's existence.

Bill arranged for me to visit Tebekwe, a medium-sized gold mine close by. When I was underground touring their workings I found a small piece of ore with a sizeable gold nugget. Since I had found it I was hoping to take it as a souvenir, but the country's law did not permit that, and I had to leave it behind.

The Rainsfords—who had survived the fifteen-year civil

Photo 72: Victoria Falls and Zambezi River, Zimbabawe

war and the terrorist activities at the time of Ian Smith's regime in the old Rhodesia, as well as President Mugabe's increasingly insane behaviour—were really great fun, and we enjoyed every minute of their company.

21

Retirement from UNSW

After returning from my fourth sabbatical leave I could detect a distinct change of atmosphere at the department. I felt that the strategies of the new Head of the mining school were linked more to obtaining funding than to promoting the welfare of the students' careers. So I decided to take retirement from the UNSW at the end of 1996. However I agreed to teach the "blasting" component of the course until a replacement was found, which in effect meant staying on as a "Fellow" for a two-year term.

I was invited by a university in Peru to present a blasting course there in 1998, and this sparked us to also visit other parts of South America, and our originally modest itinerary became quite comprehensive. Starting from Sydney we spent three days in Papeete (Tahiti). French Polynesia was quite exotic and, for me, a new experience of living on a Pacific island.

Then we went to Santiago in Chile with a brief stopover in Easter Island (a Chilean possession) where the famous and unique *moai* (carved stone statues) can be seen around the island. These are usually identified as heads only, but the *moai* are actually one-piece figures with heads and truncated torsos. (Photo 73)

The theory that is most widely accepted is that Polynesian

colonisers of the island from about AD 1000–1100 had carved them, although some are thought to have originated as long ago as AD 500. As well as representing deceased ancestors, the *moai* were erected on ceremonial sites. They might also have been regarded as the embodiment of powerful living chiefs. They were important lineage status symbols as well.

Photo 73: Stone curved heads, Easter Island, Chile

In Santiago, the capital of Chile, we stayed for two nights in a boutique hotel, and visited the Santa Rita winery in Maipo Valley. The vineyard, with its border of rose bushes, a lilypond and a fountain, was strikingly beautiful. (Photo 74)

Photo 74: Santa Rita vineyard in Maipo Valley, Chile

In Bolivia

When we flew from Santiago to La Paz, the capital of Bolivia—nearly four thousand metres above sea level—we were both affected by the altitude and felt quite ill. Brigid had to sit down, with a raging headache, and I was growing concerned about her. The situation worsened when we found out that a general strike that day meant that there would be no transport available to take us to our hotel. I begged the airport staff to arrange some kind of transport so that Brigid could lie down, for fear of her condition worsening. After some deliberation their solution was to hire a private car illegally, and showed me a car with a driver. My first shock was the condition of the car, which would not be out of place in a junkyard. The next shock was the driver, who looked like Hollywood's idea of a gangster; and the final shock was the amount of money I had to hand over upfront.

In the circumstances I had to ignore these worries, and we got in the car with our luggage. Strikers were blocking the main road, so we had to make a detour on a rough track. This was taking such a long time that I started to think we were being kidnapped—but, after an hour or so bumping along we stopped at a barricade manned by a group of strikers, where at least we could get a taxi. Struggling with our luggage to the other side of the barricade we found a taxi and continued to the hotel; I was relieved to know that we were back in civilisation and didn't have very far to go.

At the hotel Brigid didn't want to eat anything and went straight to bed. I had to prepare for my next day's visit to audit a mine's blasting practice. The hotel staff gave me something to chew (coca leaf?), which was supposed to relieve the altitude sickness. Unfortunately it didn't work, so I had an early night.

The next morning a car called for me to take me to the mine, which at that time belonged to an Australian company. The mine was located five hundred metres higher again than La Paz, which didn't help my headache or general health. In the mine workings I inspected and mentally noted their blasting practice. I was very glad to come out of the mine and shower and change back into clean clothes myself. Then came the bombshell—the mine officials wanted to hear the results of my deliberations right then and there, and not wait for my written report later; this was despite my thumping head and racing heart. I guess that I must have spent about half an hour talking to them and answering questions; I could not remember afterwards what I had said—but back in Sydney some weeks later I received an email from the mine manager saying, "Your recommendations are working like magic. Many thanks indeed." Frankly, I could not remember my recommendations; but I was certainly glad to know that they worked!

We could not get rid of the altitude sickness during our stay in La Paz, and our memory of the time spent there was not very clear. We booked our seats in buses en route to Puno, Peru, stopping first in Copacabana on the banks of the sacred lake of the Incas, Lake Titicaca, the world's highest navigable lake, on the border with Peru.

We were still suffering from altitude sickness, the main symptoms being severe headache and loss of energy; even walking a few steps was a major effort. When we got off the bus near the gate of our hotel we had to drag ourselves to reach the hotel office. The hotel staff were very pleasant, and gave us a drink which would supposedly relieve the symptoms. Tired as we were, we didn't want to waste time sleeping so we strolled around to the beach on the shore of the lake, and visited the beautiful seventeenth-century cathedral where traditionally the local people took their new vehicle to be blessed by the saint. (Photo 75)

In Peru

The next morning we boarded a bus, then a ferry across Lake Titicaca to the Peruvian side and another bus to Puno, where we were met by members of the university staff and taken to

Photo 75: (top) Titicaca Cathedral, Peru; (bottom) Cars are being blessed in the Cathedral

a hotel. We stayed there for a week, during which I was busy giving a blasting course to staff and some senior students. (Document 8)

We were simply drowning in the warm hospitality of our hosts, and at the end of our stay I was presented with a plaque and a certificate at a formal ceremony. (Photo 76)

From Puno we took a train to Cuzco, the capital of the ancient Inca Empire. We stayed there a couple of nights recuperating from altitude sickness; the

Photo 76: A plaque presented to Gour by Puno university

Document 8. Poster for Gour's one week blasting course at Puno university, Peru

elevation of Cuzco is about five hundred metres below that of Puno, so there was some relief. Then we went by train to Aguas Calientes, where we stayed two nights in a hotel.

Anyone energetic and fit enough follows the Inca Trail from Cuzco to Machu Picchu, taking four days to reach it, staying the three nights in staging camps. All the structures there were built as drystone walls, using huge blocks of basalt cut and smoothed with such precision that when they were placed in position there is no visible gap. How they managed to accomplish this fine art is still a mystery. (Photo 77)

Photo 77: Ruin of a burial tomb, Machu Picchu, Peru.
Note the fine joints between stone slabs

We went by bus from Aguas Calientes to Machu Picchu, but afterwards Brigid was not feeling well and she decided to stay in bed. What should I do? I decided to experience the last section of the Inca Trail, from Aguas Calientes to Machu Picchu, which climbs about six hundred metres. The steps were about half a metre high, quite a lot higher than we normally encounter. My plan was to return by bus, but after arriving at the top I found that the next bus would not be leaving for another three hours. So, again, what should I do?

I decided to walk back, which was a stupid option to take. By about half-way down I realised my mistake. Going downhill, the high steps caused continual impact on my knees; however, this was the point of no return! So I kept going down, and when I reached the bottom of the trail my knees collapsed, and I had to sit down for a while. After some rest I managed to walk to the hotel using what energy I had left. Obviously, I must have damaged my knees, a condition that flared up later.

In Columbia

After returning to Cuzco we flew to Bogotá, the capital of Colombia, with a short stopover on the way in Quito, Ecuador, the usual jumping-off point for the famous Galapagos odyssey. From Bogotá we flew to Cartagena to meet our third sponsored child through Plan International. It was a very valuable experience to meet the girl face to face and to find out how our donations were used. One of the important initiatives taken by the local Plan office was to spend some of the sponsored money on building materials to be used to build a solid homes with voluntary help from the community together with those shanty dwellers who had signed up for the scheme.

In that part of Colombia there were no security concerns, and we were able to walk freely along the streets, even at night, without fear. We took that opportunity to see the local market in action as well as the five-hundred-year-old Spanish fort built to protect the city from invaders.

From Cartagena we flew to Buenos Aires, originally with the idea of visiting Patagonia. This turned out to be impossible to organise within the short time available to us in Argentina before flying back to Sydney. We made the most our time sightseeing in and around Argentina.

In 1999 I was invited to present a seminar at the Technische Universität, Berlin, and we visited friends in England, France, Denmark, Austria, Germany and Hungary.

We also included in this trip some other countries, such as Switzerland, that we had not visited before. Before leaving Sydney, we bought Eurail first-class passes for train travel to most of those places.

An Unsavoury Experience

An interesting incident occurred on one leg of this trip. In Vienna we validated our Eurail passes for travel from Budapest to Berlin and bought our tickets, boarded the train in Budapest and made ourselves comfortable.

When, a few hours later a ticket inspector checked our tickets, he told us they were not valid. His spoken English was very poor but we gathered that the train was going through Slovakia, which was not part of the Eurail group: cash only for the Slovakian leg of the journey. He would not listen to our pleas of innocence. We offered to pay by credit card but this was refused. I even showed him the document stating that we must be in Berlin tonight as I had the seminar there the next morning. The inspector was adamant that only cash in Slovakian or German currency would be accepted. There was an ATM (one of only two in Slovakia at that time) at the railway station; but the inspector would not hold the train for me to get the money from that machine. By this time another inspector had joined him, and they were extremely rude in their behaviour: since we did not have the required cash we were forcibly ejected from the compartment onto the platform in Bratislava.

Standing on a deserted platform in an unknown country on a Sunday morning was a very low point in our travels. My first thought came in my mind was, "Why do we travel so much instead of having a cosy, stable life at home?" We had to do something about this mess. In the station, Brigid spotted a noticeboard showing times of trains in and out of Bratislava and saw that there was an overnight train for Berlin leaving at around ten p.m. If we caught that train, it would mean that I would be in Berlin for my seminar, although somewhat late. I took out a large amount of local currency from the ATM and bought tickets to Berlin. Then Brigid, who is fluent in German, phoned Dr.-Ing. Helmut Wolff at the university in Berlin and explained our situation. He said that someone would meet our train in Berlin and take us straight to the university.

At least the seminar in Berlin would go ahead, much to my relief. Then the question was how we should spend the day here, now that we had plenty of local currency. We took a tram from the station to the centre of the city and had our lunch there—surprisingly cheap—then took another tram to the bank of the Danube River, and there visited a museum. We met an army officer who spoke some German, and through a combination of speech and sign language we discovered the existence of a wetlands bird sanctuary, and what bus to catch to get there.

We were pleasantly surprised at how beautiful the wetlands were. We spent a few hours there exploring the place and enjoyed it very much. We returned to Bratislava well before dark and caught our overnight train for Berlin. Although the initial drama of being kicked off the train was very traumatic, in the end we had the benefit of getting an unscheduled glimpse of Bratislava.

Helmut Wolff's daughter met us at Berlin railway station. She took me straight to the university and to a room full of delegates patiently awaiting my arrival; the presentation went well and I was very pleased with the end result.

We were booked into our hotel for three nights in order to do some sightseeing in and around Berlin. Professor Wolff very kindly escorted us to all the tourist sights, including one of great historical interest: Cecilienhof in Potsdam— originally the home of Crown Prince Wilhelm Hohenzollern —where the Potsdam Conference was held in 1945 between the leaders of Great Britain, USA and USSR to discuss the fate of the defeated Germany. (Photo 78)

Photo 78: (top) Potsdam Conference between Great Britain, Soviet Union and USA; (bottom) the venue of the meeting

22

New Paths

Brigid and I thought that we should now do some volunteer work to serve the wider community, and we enrolled with Australian Volunteers International (AVI), which was mainly funded by AusAID, the Australian Government Overseas Aid Program.

After that I joined the University of Wollongong primarily to conduct research work as an Honorary Visiting Fellow in the Engineering Faculty. A postgraduate student, Les Armstrong, was completing a doctorate in my area of special interest and needed a supervisor, so the arrangement suited all parties. I was also invited to present papers at conferences and give seminars to staff and students.

While I was reading a newspaper one day Brigid took a long phone call. When she finished she came and announced that the call was from someone at AVI inviting her to teach for a year at a Chinese university.

My immediate response was, "China! Not again!" (Our past experience was still very fresh.)

After some discussion we decided to follow that enquiry through. It appeared that Brigid was designated to teach English at the Central South University (CSU) in Changsha (Hunan Province). AVI could place me there as well, because CSU taught a mining engineering course.

It appeared that a group of CSU academics had visited UNSW a few years earlier when I was the head of Mining Engineering. I had hosted the group and arranged their mine visits, among other events, and now I was being invited to accompany Brigid for a year, beginning in July 2000. In such volunteer arrangements, AVI paid return airfares and travel insurance, and our boarding and tuition expenses for an intensive three-week course in basic Chinese language. The host university would provide western-style furnished accommodation and would pay a nominal salary to cover our living expenses.

China Assignment

After a two-day orientation course in Melbourne with other volunteers, our group of five travelled to Shanghai. A man holding up a big signboard saying "Professor Sen Etc." met us at the airport. This became a standard joke between us— when we had to refer to any other member of our group we would call, "You etc!"

From Shanghai we travelled some distance by train after crossing Yangtze River to Nantong for our rudimentary Chinese (Mandarin) language course. I found learning Chinese very difficult. In fact the teacher divided us into two groups: I was put onto the "slow" group, whereas Brigid was in the "fast" group. My group grasped some of the most important vocabulary, but even then I wasn't confident of the correct pronunciation. (Photo 79)

Central South University (CSU), Changsha

After suffering the torture (for me, at least) of trying to learn Chinese, we were all dispersed to our allocated universities. In our case, Brigid and I travelled to our university in Changsha. (Photo 80)

Photo 79: Chinege language training course for Australian volunteers, Nantong, China

Photo 80: CSU campus where Brigid and Gour taught for 18 months, Changsha, China

When we arrived at the campus we met Ms Zhang, our "minder", who looked after overseas staff at CSU. (Photo 81)

Photo 81: Gour being a pillion passenger when Ms Zhang (Inter. Coordinator) cycling in CSU campus

We were then taken to our flat, which had been hastily converted from an Asian to a Western style: as you stepped in the door, you were in the dining room, where there was a fridge. To the left was a tiny bathroom, too small to hold two people at the same time, with a shower and its hot-water tank, and a toilet; at the front was a small kitchen, and to the right, a bedroom with a small air conditioner and a sitting room with a desk and a couple of chairs. There was also a telephone in the flat.

Hardly any linen was provided; we had to immediately buy some reasonably good sheets and towels. We had the option of cooking our own meals or eating at the students' canteen, which we did once a week. There were restaurants outside the campus but they were so far away that we were reluctant to go there.

In every class at the university, a monitor had been selected by the Communist Party authorities to report any objectionable incident. All first-year students underwent military drill for a period on the university campus. (Photo 82)

Photo 82: CSU students undergoing military training in the campus

In the meantime, the authorities discovered that I had hosted that group of CSU academics visiting Australia. That altered our positions—my status was raised from being Brigid's 'Dependent Spouse; suddenly I was treated as if I were the more important. The university top management arranged a banquet for us at which I was the guest of honour, and I was installed in a well-decorated office with silk curtains, a phone and even a computer—all much to Brigid's indignation.

Living in China was difficult at first, with the language problem, the alien environment and complete lack of most Western-style food; however, we gradually became used to the surroundings and accepted the various shortcomings of living there. As winter approached we really felt the inadequate heating arrangements in our flat, and asked for more heaters. The authorities invented excuses ranging from, "The university cannot afford to pay for running the extra heating," to, "The electrical wiring won't cope with an extra load"; and on and on. Our trump card, however, was to point out that it was impossible to prepare our lessons in a cold room, so we needed a separate heater in the sitting room. Eventually they supplied an extra heater on the condition that we would not have two heaters turned on at the same time (or, presumably, they would remove one of the heaters). Without adding this proviso, their having to agree with us would have been a serious loss of face for the authorities, the superior giving way to an inferior—another example of the Chinese way: they must not be put into a corner; there must be an exit strategy. (As a result, the other expatriate teachers—two Canadians, two Japanese and one American-Chinese—were able to enjoy the luxury of being provided with adequate heating.)

In the university's winter break in January we visited my family in Calcutta, via Hong Kong. This was a very welcome break from our rather isolated life on the CSU campus, which was more than four kilometres from Changsha city centre. The distance was not great but the narrow road was rough and highly congested with traffic. Taxis were difficult to get, and a rather infrequent bus service was the main mode of transport. The main attraction for us in the city was the Western-style Heiwado department store that, among other things, boasted an authentic Brazilian *churrascaria* restaurant, which we visited several times.

****

One day one of my Chinese colleagues asked me, "Do you know Professor Dexin Ding from another university? He heard that you were here, and wanted to see you."

Obviously my presence in China must have been passed along the grapevine. I said, "I knew a Chinese academic called Mr Ding some years back. He must have been promoted to professor now. Yes, I shall be very happy to meet him."

He said, "When dignitaries from other universities visit us, our custom is to give a banquet. You and your wife are invited."

Soon after we returned to our flat after our classes on the day of the banquet there was phone call from my colleague professor saying, "Professor Ding wants to see you in your flat before the banquet." I asked him to delay the visit for at least fifteen minutes so that we could tidy up the flat before he came.

When he came he brought two bottles of *moutai*, the famous—and very expensive—Chinese liquor. I was somewhat taken aback by his generosity, but it was very nice to meet him again after eight years or more.

After the banquet he invited me to give a seminar at his university about three hours away by car, which he would arrange. This meant that I would have to stay overnight at a hotel there, and return in their car the following day. Brigid was also invited but her schedule wouldn't permit her to take the day off.

When I arrived at the university, I was met by a group of dignitaries, including the equivalent of the Vice-chancellor, who welcomed me warmly. (Photo 83)

Photo 83: Professor Ding (far left) and other high officials of Nanhua University with Gour

The seminar was to be held in a large hall, and was filled with students and staff. Afterwards Professor Ding gave me an envelope that I thought would be a thank you letter, so I didn't open it at the time. (When I opened it later at home it contained a thousand Yuan, which then was roughly what a university teacher would earn in a month.)

In the evening after the seminar a banquet was held, which all the dignitaries attended. Professor Ding was sitting next to me, and I wanted to square up our earlier encounter in Sydney. So I said, "Do you remember the time I used some strong language to you when you were in Sydney? I'm sorry for that."

After a pause he said, "Yes I do! That was good for me." I was very relieved to hear it.

I was also invited to present another seminar there, which I fully intended to do, but in the event other commitments prevented it.

****

During my stay at CSU I visited some collieries and metal mines, and was asked to audit their blasting practice in the hope of helping to improve their safety and productivity.

One such invitation came from Tongling copper mine in Anhui Province. (Document 9)

They offered to pay the air fares and hotel expenses for Brigid and me, and provide an interpreter. We were met at the airport by their car, and were taken straight to the mine office. There, a group of senior mine officials warmly greeted us. After that, however, there was no mention of our going down the mine; instead, their blasting practice was described to me with the aid of printed diagrams and by drawing sketches. When I said that I really could not comment until I witnessed the actual operations at the site, there was an uncomfortable silence. Apparently they had discovered from my passport (which I had been obliged to surrender to the host institution, at that time) that I was over seventy years of

Invitation

To: the Central South University

Attn: Mr. Gour C. Sen

Date: May 10, 2001

Subject: Visit to Tongling

Dear Mr. Gour C. Sen:

We highly appreciate the achievements you have acquired in mining world. We are glad to know that you are now in the Central South University. We hope you could come to our company for a visit and give your direction to us in mining production.

We are waiting for your reply eagerly.

With kind regards

Sincerely yours

Tongling Nonferrous Metals (Group) Inc.

Vice President: Wang Shanyuan

Document 9: An invitation from Tongling Mine in China to audit their mine

age, and for safety reasons no-one that old was allowed to go underground in that particular mine. I pointed out that I had visited other mines in China previously, but that made no difference.

After that I tried to make some comments about their blasting practice, which they promptly wrote in their notebooks. However I was not satisfied with the outcome, and felt that the mining company had largely wasted their money on that visit.

'Rescue' Mission

One evening I had a phone call from Glenda Lasslett, the then Regional Manager of AVI in China asking if I could help a young lady—also an AVI volunteer who was teaching at a university in Changsha some ten kilometres from CSU—with her return travel to Sydney from China.

The story was that the young lady hadn't realised that her visa had expired. This was a serious offence, and the police were threatening to arrest her, and she was afraid of being on her own away from the university campus. When the AVI in Australia heard about this, they promptly organised flights for her from Changsha to Sydney via Hong Kong. I was asked if I could meet her at her university and escort her through the check-in counter at Changsha Airport.

I had to go by taxi at half-past five the next morning, so I thought that I had better ask my Chinese colleague, Professor Wu Chao, to teach me the exact pronunciation of the name of the university where I had to go. I mimicked that word several times but I wasn't fully confident. So I went to Wu's apartment that night and asked him to write the name in Chinese in my diary. In the morning I went to the equivalent of a taxi rank. When the driver asked in Chinese for my destination, I thought that I should try my pronunciation; but it was no good. I could see that if I kept trying for much longer, he would drive away in frustration; but when I showed him the written version, he was all smiles, and I hopped in the cab.

The operation went very smoothly, although I was a little anxious. Fortunately the police didn't intervene. I waited at the airport until her flight left, feeling much relieved.

****

Near the end of our one-year contract with CSU the university authorities approached us with the invitation to extend our stay in Changsha for further year. By then we were fairly well adjusted to the environment in China and, after due consideration we decided to compromise and extend our contract for six months, but not for a year. Generally Chinese universities had two long vacations each year, one in summer (June/July) and the other in winter (January/February), the winter break being the longer. During these breaks we were able either to see more of China, or get away from China altogether.

By the end of our extended contract we had accumulated a lot of personal belongings that we were loath to leave behind, so we decided to send some to Sydney and some to Stroud in England. No overseas removal companies operated in Changsha at that time, so everything had to be mailed in cartons of a certain size and weighing not a gram over twenty kilograms. The cartons must have nothing written on them except the address; each one must be filled in front of a post office employee, who then weighed the carton on electronic scales. Altogether, we had eleven cartons to send overseas.

On the day before our departure in January, 2002, we were very busy packing, when a large limousine pulled up near our building, and Professor Ding arrived unannounced. We invited him to have a drink with us but he politely declined and said, "I know you are very busy packing. When I heard that you are leaving I couldn't help but come over to say goodbye." He shook hands with us and went, leaving a small packet as a gift. We were dumbfounded—he was spending a good six hours travelling for a five-minutes visit; we couldn't help but salute him. We had no time to open the packet, so it went into our baggage. When we opened it in Sydney we were delighted to find a large silk embroidery which depicting eight flying cranes against a woodland backdrop. (Photo 84)

Photo 84: This silk embroidery was presented by Prof. Ding to Gour just before their departure from CSU

We felt sad at the time to be leaving CSU. In eighteen months we had met a host of various trades people—handyman, shoe repairer, baker, tailor, greengrocer. Despite the language problems we had connected with them quite well.

****

In Tibet

Before leaving China we visited Lhasa in Tibet—at 3,490 metres above sea level, one of the highest cities in the world, often called the Roof of the World. Special permission was required to go there, and it was stipulated that an authorised travel agent in Lhasa must arrange our itinerary and, when travelling outside Lhasa our travel agent must accompany us. Fortunately our travel agent was a Tibetan, not Chinese, and this ensured a relaxed relationship with us. On one occasion we visited Drepung Monastery, the great Buddhist centre of learning established in 1416; it is about seven kilometres from the centre of Lhasa. (Photo 85)

Photo 85: Drepung Monastery, Tibet

One of the interesting features of the Buddhist monasteries was the debate between the monks, a traditional form of theological exercise that tourists are welcome to watch. (Photo 86)

Potala Palace, the traditional residence of the Dalai Lama

Photo 86: Monks are debating practice in Sara Monastery, Tibet

from the seventeenth century until the Chinese invasion of Tibet on 15 March, 1959, and now a museum—thirteen storeys high, with over a thousand rooms—is indeed an extraordinary landmark in Lhasa. Its majestic appearance on the mountain slope will astonish any visitor to Lhasa. (Photo 87, Photo 88)

Photo 87: Dalai Lama's official residence, Potala Palace, Lhasa, Tibet

**Photo 88: Dalai Lama's Winter Palace;
Dalai Lama fled from here to India**

We stayed in Lhasa for five days, during which time we visited the Winter Palace where the present Dalai Lama lived before he escaped, disguised as a Chinese soldier. He and twenty of his entourage travelled at night to avoid the Chinese sentry guards for fifteen days over the Himalayas. He had to cross the five-hundred-meter-wide Brahmaputra River and endure the harsh climate and extreme altitudes, finally crossing the Indian border at the Khenzimana Pass; he has lived in India ever since. Many thought he had died in the fierce Chinese crackdown that followed the Tibetan uprising.

After our visit to Tibet we had a short holiday in Japan before heading back to Sydney.

PART 6

Retirement Period

23

Semi-retired Life

After returning to Sydney my time was spent in occasional consulting work for the industry, going to Wollongong University (UOW) once a week to supervise postgraduate research and writing papers for technical journals and conferences.

In November 2002 our first grandchild, Hugo, was born in London to Jane (previously called Runa), our middle daughter. Fortunately we were able to visit the UK a few months later and saw Hugo as a baby in their country house, but unfortunately London and Sydney are too far apart for us to see Hugo regularly. (Photo 89)

Photo 89: (top) Family gathering, [l to r) Rahul, Jane, Anita, Tony (Jane's husband), Shuki with Hugo, Gour and Brigid. (bottom) Jane & Tony's mansion in Surrey, England

Our two bed roomed flat in The Spot in Randwick was rather cramped—and it required us to climb fifty-six steps. We were fortunate to find a larger apartment just down the hill from The Spot, in Coogee, which had only ten steps to the front door – and so we moved.

Unfortunately the housing market slumped

Photo 90: **Happy Gour holding his first grandchild Hugo**

suddenly around that time, so we could not sell our flat in The Spot, and had to let our bigger flat in Coogee. In 2004, after the tenants had vacated the Coogee flat, we moved back in, having managed at last to sell in The Spot, and were happy to have more room to spread out. The tenants hadn't looked after the apartment properly and we had to make repairs and renovations. My daily exercise walking to Coogee Beach was very handy, and the seawater was also very soothing.

Eventually I received a letter of appointment from UOW which stated, in part: "Your appointment entitles you to the use of the title of Professor and to similar academic standing, privileges and responsibilities as for other Professors at the University". It was a welcome recognition from UOW.

I had always had a soft spot in my heart since childhood for plants and flowers, going back to the days in our house in Dehri on Sone with its large garden. That affinity flourished in Elm Tree Cottage in England, where there was a sizeable garden to look after. After we migrated to Australia I had been seriously involved in gardening whenever we visited our cottage in Stroud.

In Australia we have always lived in an apartments. Although I did not have any access to land, I grew flowers on the balcony, particularly orchids. Then in the first decade of the new century I joined an organic community garden where I worked with the plants each weekend, but that ended when the land was needed for some other purpose. (Document 10)

Document 10: Gour's involvement with community organic gardening (shown within the drawn circle)

Our apartment is one of six in a block. The grounds have a number of mature shrubs, a large lawn and a neglected plot of ground that had been used as the builders' dumping ground, and was harbouring weeds and other noxious plants which had to be destroyed professionally every couple of years, paid for by the body corporate (i.e. the group of owners of the block of flats). At a body corporate meeting I proposed that I transform it into a vegetable garden for organically grown herbs and salad plants for our use—mine and Brigid's—in return for my looking after the shrubs; up till then the cost of pruning the shrubs had been quite substantial. With their approval, I began clearing the builders' debris and, over time, filled the plot with fresh soil and compost and planted seeds and plants, all organically grown, using no chemical pesticides or fertilisers. Gradually the garden produced some good vegetables, which I often shared with other residents in the block.

In 2004 our local council, Randwick City Council, held their first Ecoliving Fair promoting organic food and a healthy lifestyle. I won a large compost bin in a competition that involved completing a questionnaire, and have used it for the garden ever since.

****

In 2004 Rahul, our doctor son, married Jenny, also a doctor, in the beautiful setting of Lizard Island on the Great Barrier Reef. It was a momentous occasion for Brigid and me, being the only wedding of any of our children that we had attended. (Photo 91)

Soon afterwards they delighted us with our second grandson, Oliver. (Photo 92)

They settled in Sydney, so we see them very often. Oliver stays with us one night a week, and it is always a delight for us. Ever since he could talk, Oliver has called Brigid "Ba", and still does now, even beyond the age of seven.

****

Photo 91: Brigid
and Gour at
Jenny and
Rahul's wedding
in Lizard Island,
Queensland,
Australia

Photo 92: Proud Grandfather holding baby Oliver

Golden Wedding Anniversary

The year 2005 was also a milestone for us—our golden
wedding anniversary, which we celebrated in three places:
in Sydney, in Calcutta and in Stroud. We went back to Roath
Park in Cardiff to have our photograph taken, the spot where
we had been photographed once before—in 1954, not long
before we were married, then again in 2005. It is interesting
to see how we changed over that time. (Photo 93, comparing

Photo 93: In 2005 B and G on the same bridge in Roath Park, Cardiff as in 1954 (compare Photo 13)

this with the photo 13 in Chapter 2: Cardiff)

In the course of a year in the UK we refurbished our cottage in Stroud and visited many friends and relatives. In the meantime Oliver had grown, and was now three years old. Once, when I was looking after Oliver on my own, he tried to say something to me. I didn't understand him and asked him to repeat it; he took my hand and guided me to a cupboard, opened a drawer and, without a word, pulled out my hearing aids from the drawer and handed them to me. What a smart boy—how did he know about my hearing problem or what the hearing aids were for? When I put the gadgets into my years he repeated his words. His talent for grasping the experiences of life has flourished as he has grown.

****

In 2007, UNSW Press informed me that the copies of my book would be sold out within a year, but they were no longer publishing textbooks by then. It was more than fifteen years since I had written the original manuscript, and some parts required revision. I decided to write a revised, expanded edition that would attract a wider audience.

It took me more than a year to compile the draft, which I sent to EA Books, the publishing arm of the Institution of Engineers, Australia, for their consideration. Their reviewer's comments were encouraging, but it required a major re-shuffle of the contents. Initially EA Books wanted to publish the book in digital format but in the end they agreed to publish it in both printed and digital formats.

Meanwhile I was aware that Imperial College, London, was about to get funding for a very large research project looking into the cause of an Australian mine accident in 1999 (Northparkes Mine, Cobar) in which there were four fatalities from windblast caused by falling rock. After approaching Imperial College they invited me to spend a year there, commencing in June 2008. (Document 11)

Imperial College, London

We set off for London in May, 2008, via Japan, where we spent about three weeks, mostly in Tokyo and Kyushu. After arriving in London I used the library facilities at IC to complete the revised manuscript. I was very impressed by their research facilities and the standard of the academic staff; however, I could not become seriously involved with the Northparkes project as I wasn't a staff member: the confidentiality agreement between the mine owners, Rio Tinto, and Imperial College precluded my being informed of the project details.

This news was very disappointing, as I was mentally fired-up and ready to be involved in the project. Our whole plan was somewhat derailed. We could not go back home, as our apartment was occupied by tenants for the whole year, so we had to redraw our plans to utilise our time away from Australia profitably. Besides seeing friends and relatives more

Imperial College
London

Department of Earth Science and Engineering
Imperial College London

Room 1.35, RSM Building
Prince Consort Road
London SW7 2BP
Tel: +44 (0) 20 7594 7327
Fax: +44 (0) 20 7594 7444

j.p.latham@imperial.ac.uk
www.imperial.ac.uk

05 May 2008

John-Paul Latham BSc, MSc, PhD, CGeol

Professor Gour C Sen

2/16 Carr Street

Coogee, 2034

AUSTRALIA

TO WHOM IT MAY CONCERN

Dear Sir/Madam

This letter is to confirm the Prof Gour Sen has been invited to Imperial College London as an Academic Visitor by Dr J-P Latham for a period of one year. He is expected to arrive in London on June 8th 2008. He will be an academic visitor to the Department of Earth Science and Engineering.

Yours faithfully

J.-P. Latham

Document 11: Invitation from Imperial College to spend one year as a Visiting Academic

often, we decided to do some major improvements in our cottage.

Unknown to me, in 2008 Brigid and our children had secretly organised a huge party at our cottage in Stroud, with a large marquee, to celebrate my eightieth birthday in September. With caterers to do all the work, we were able to relax completely with almost seventy of our relatives and friends. Many had travelled long distances and some had come from overseas. (Photo 94)

Photo 94: Gour's 80th Birthday lunch at Elm Tree Cottage, Stroud, England (2008)

A Lucky Escape

I went several times to London to have discussions at Imperial College, but we mostly stayed in Elm Tree Cottage in Stroud, with its large garden and fruit trees. One day Brigid was out shopping while I stayed home to let in a tradesman, Alan Harrison, to do some painting. I was outside picking gooseberries from the garden when a wasp stung me. I ignored it, but suddenly a swarm of ten wasps or more attacked me and stung my face. I began to feel drowsy and headed indoors, shouting to Alan to tell him what had happened, and collapsed on the stairs. Next thing I remembered was being attended by two ambulance men, one holding an oxygen mask to my face and the other giving me an injection.

Afterwards I was told that Alan had called 999 for an ambulance. When I become fairly stable I was driven to hospital for further checks and medication. I learned that I was now allergic to the stings of both bees and wasps, and should always carry self-injecting antihistamine medication as a safety measure. On reflection, had he not been in the cottage I would have been unconscious for a long time, which could have been fatal—another escape from death?

EA Books published the revised and enlarged edition of my book shortly before we left England for Sydney in April, 2009. (Photo 95)

**Photo 95: Gour's second book published by
Engineers Australia in 2009**

In March 2009 we heard from Masahiro (Mike) Miyazaki, who had been a friend of our late son Prasanta, that Fumiko, Prasanta's ex-girlfriend had died. After Prasanta's death

Fumiko had adopted us as her own parents. We became so attached to each other that whenever we visited Japan we spent a few days with her, and she had also stayed with us a couple of times in Sydney. Due to various commitments in England at that time we weren't able to attend her funeral.

****

After returning to Australia I learnt that one of the mining engineering teaching staff at UOW had retired, and the university was trying to spread the retiree's load onto the rest of the staff. With that in view I was approached to take some of the teaching load. I felt, however, that I was content being involved in postgraduate research work, and did not want to take on undergraduate teaching. After lengthy discussions with the university I decided to sever my ties with them.

My first reaction was to write another book that would assist future researchers in the field of blasting. I knew that it would be published in digital format so that it could be updated every three to five years, and at my advanced age it seemed wiser to have a younger co-writer who could carry on the work after I could no longer do it. I approached several people, among whom was Dr Les Armstrong, one of my ex-postgraduate students, who agreed to take on the role. He made some initial preparations for the book but had to pull out because of health and other problems. At that point I decided not to pursue it any further but to document my own life's experiences instead. My grand-niece, Adrija, also encouraged me to write my memoirs; thus inspired, I began preparing *Major Miner*.

In March, 2009, we received a message from Mike Miyazaki in Japan that the first anniversary of Fumiko's death was to be held in a ceremony at the Zo-jo-ji Temple in Tokyo where Prasanta's ashes are kept. Since we did not attend Fumiko's funeral, we wanted to be at the first anniversary ceremony of her death, which is very important in Japanese society.

It was a sad as well as fruitful occasion, where we met some more of Prasanta's friends. One of our Japanese friends, Professor Sohei Shimada from Tokyo University, invited us once more to stay in his modern apartment in Tokyo. (He played the double bass, and had three instruments that he kept at the apartment because his own home wasn't large enough to take them.) As always, Sohei and his wife Yoko were most hospitable during our stay in Tokyo.

Back in Sydney, I kept up my hobby garden, probably spending even more time and energy on it than before. Unfortunately insect pests destroyed most of my orchid plants—although, paradoxically, some discarded plants that I tied to an olive tree in the front of our building have since produced some very good flowers. I still attended orchid society monthly meetings to see the serious growers' achievements.

Since we moved to our apartment in Coogee, after a time I made friends with the wild Rainbow Lorikeets and fed them sunflower seeds, their favourite food. A couple of them had become so used to me that they would fly over the balcony and cling to the window to attract my attention; or, when I sat on the balcony, they would perch on me. (Photo 96)

Photo 96: In Sen's balcony wild lorikeets all over Gour demanding food!

Presented 4 lectures (nearly 3 hrs) at 82yrs!

Occasionally I have been invited to give lectures both in Australia and overseas. One such recent assignment was a series of four lectures at Chiang Mai University in northern Thailand during the fortieth anniversary (December 2010) of their engineering faculty. (Document 12)

I presented all four lectures in one marathon effort, taking a little over three hours, with a fifteen-minute coffee break. That was a record performance for me, especially considering that by then I was eighty-two years old. After leaving Chiang Mai we spent a few days in Chiang Rai in the northern mountain range of Thailand. One excursion from there took us to the Golden Triangle, which is the strategic border between Burma, Laos and Thailand—and possibly most famous as an illegal opium-growing area in Asia. (Photo 97)

After our visit in Thailand we visited our Plan International sponsored child in Dhaka, Bangladesh, where we also met the relatives of our daughter-in-law Jenny, including her mother. We were somewhat shocked at the chaotic traffic on the roads: on one occasion it took us forty minutes to travel three kilometres.

Photo 97: Golden Triangle between Burma, Thailand and Laos

INSTITUTE OF EXPLOSIVES ENGINEERS
AUSTRALIAN BRANCH

PO BOX 3198, MARRICKVILLE METRO, NSW 2204

Tel: 0414 342033

Patron: Professor Gour Sen: PhD, CPEng,
FIEAust, HonFIExpE

AUSTRALIAN BRANCH ANNUAL REPORT

FOR THE 2011 ANNUAL GENERAL MEETING

1. Elected at the Australian Branch AGM in November 2010, the following members will comprise the management committee for 2011. Mr Kevin Wall as Chairman, Mr Dick Godson as Vice Chairman, Mr Brian Roberts as Secretary and Mr Alan Hind as Treasurer. The full financial accounts as presented to the AGM have been provided to the Secretariat of the Institute. The Australian Branch is in a healthy financial position with about $A28,000 held in term deposits and about $A5,000 in the working account. The ability of Australia-based members to pay their annual subscription through the Australian Branch continues to provide a convenient and economic process that assists the members and the Australian Branch.

2. A draft set of Branch Rules have been developed and provided to the Council for review and ratification.

3. The Australian Branch currently has 55 members, comprising one Honorary Fellow, 3 Retired, one Associate and 50 Members. While this is a reasonable number for the Branch, and membership covers a range of occupations, the future plans for the Branch will involve increasing and diversifying the membership. It is planned to survey the current membership to obtain a more complete picture of the range of knowledge and skills that our members possess. One aim is to create vocational groups within the total membership that can help to overcome some of the problems associated with the large distances that divide the members. Another aim is to establish the framework for an Explosives Seminar that will be attractive, and useful, to as many members as possible.

Alan Hind

11 April 2011

Document 13

Visit to Makaibari Tea Estates

Our next visit was to India, primarily to see our relatives in Kolkata and Bangalore. While we were there we took a trip to the Darjeeling district in the foothills of the Himalayas. We had arranged beforehand to visit the famous Makaibari Tea Estates managed by the fourth-generation owner, Rajah Banerjee. The estate, covering some 637 hectares, was bequeathed by the British Raj in 1859 to Rajah's ancestor. Now Rajah manages the land, of which 223 hectares are used for tea planting and the villages housing the people involved directly or indirectly with producing the tea, and the rest is sub-tropical rain forest inhabited by wild animals, including tigers.

The tea production is all organic; no pesticides are used. To my knowledge, it was the first tea production enterprise of that type in the world. One of their products, Silver Tips Imperial, which is harvested in and around the full moon, was the world's most expensive tea at the time of our visit. While we were touring the tea processing equipment, we met Rajah Banerjee and his wife Srirupa. We tasted eight grades of tea, making a personal evaluation; afterwards the Banerjees invited us for dinner that evening at their house. (Photo 98a and b)

In their sitting room there was a great deal of ancestral memorabilia

Photo 98a: Brigid and Gour in Rajah's house, Makaibari Tea Estates, Kurseong, Himalayas

and stuffed wild animals, including a man-eating tiger—
although normally tigers are not hunted on Rajah's land, we
were told.

We were staying at the Cochrane Place Hotel in Kurseong,
a restored and refurbished nineteenth-century British
government officer's residence, from where on a clear
morning we could see the peak of snow-covered
Kangchenjunga rising to five and a half kilometres above
sea level.

I have continued my close relationship with the
Australian Branch of the Institute of Explosives Engineers.
(Document 13)

With all these activities, and walking daily to Coogee
Beach, my days were passing without any serious drama.
The best feeling at this stage of my life has been that I can do
whatever I like without being accountable to anyone except
Brigid!

**Photo 98b: (l to r) Gour, Brigid, Rajah; Rajah's Bungalow, Makaibari
Tea Estates, Kurseong, Himalayas**

24

Afterwards ...

Looking back to my early teenage days when I was behaving wildly without any goal in mind, I realise that it could have continued forever had I not become interested in rowing as a sport. My win in the Sculling event of the Inter-College Regatta was a turning point in my life, and provided me with confidence. This was followed by my association with a friend who was a catalyst for my reformed way of looking at life.

I now realise that a string of positive events has led me onto a straight and meaningful path. This would not have happened had not a determined driving force pushed me to it. The real crunch was to prepare myself to deal with any event that might threaten to derail my ambitions.

Life is full of obstacles that must be overcome by sheer determination. I was lucky enough to be in a situation where I could well have been diverted from my original goal, but I met the right people to help me cross that barrier.

In spite of a comfortable life in Calcutta I decided to relinquish a seemingly stable position. I jumped into an uncertain future in order to protect the wellbeing of the children. In the course of looking for a job I realised that I was over-qualified for a position in the industry, but I did not mind taking on a fairly low-key site engineer's position when it was offered to me. By sticking to that line of work I increasingly gained the company's confidence to the point where, at the end, the company's managerial position was offered to me.

I was lucky enough to engage myself in teaching and engineering work simultaneously in order to balance my family life and industrial commitments. This again was useful in my later consulting work when key industry personnel placed their confidence in me.

My final word is that, in order to be successful, you have to believe in yourself and believe in what you are doing. If you are also lucky enough to have an adorable and loving life's partner to support and encourage you, the task is made that much more enjoyable.